Entropy and Multivariable Interpolation

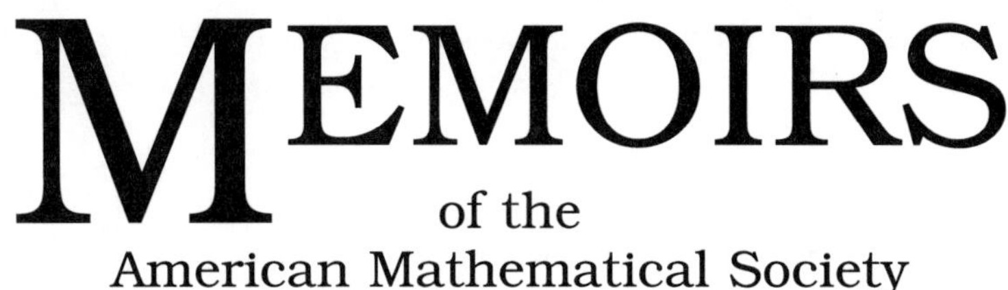

Number 868

Entropy and Multivariable Interpolation

Gelu Popescu

November 2006 • Volume 184 • Number 868 (end of volume) • ISSN 0065-9266

American Mathematical Society
Providence, Rhode Island

2000 *Mathematics Subject Classification.*
Primary 47A57, 47A13; Secondary 47A56, 47A20, 47B35.

Library of Congress Cataloging-in-Publication Data

Popescu, Gelu, 1958–
Entropy and multivariable interpolation / Gelu Popescu.
p. cm. — (Memoirs of the American Mathematical Society, ISSN 0065-9266 ; no. 868)
"Volume 184, number 868 (end of volume)."
Includes bibliographical references.
ISBN-13: 978-0-8218-3912-6 (alk. paper)
1. Interpolation. 2. Linear operators. 3. Maximum entropy method. 4. Operator theory. I. Title.

QA281.P65 2006
511′.42—dc22 2006043003

Memoirs of the American Mathematical Society

This journal is devoted entirely to research in pure and applied mathematics.

Subscription information. The 2006 subscription begins with volume 179 and consists of six mailings, each containing one or more numbers. Subscription prices for 2006 are US$624 list, US$499 institutional member. A late charge of 10% of the subscription price will be imposed on orders received from nonmembers after January 1 of the subscription year. Subscribers outside the United States and India must pay a postage surcharge of US$31; subscribers in India must pay a postage surcharge of US$43. Expedited delivery to destinations in North America US$35; elsewhere US$130. Each number may be ordered separately; *please specify number* when ordering an individual number. For prices and titles of recently released numbers, see the New Publications sections of the *Notices of the American Mathematical Society.*

Back number information. For back issues see the *AMS Catalog of Publications.*

Subscriptions and orders should be addressed to the American Mathematical Society, P. O. Box 845904, Boston, MA 02284-5904, USA. *All orders must be accompanied by payment.* Other correspondence should be addressed to 201 Charles Street, Providence, RI 02904-2294, USA.

Memoirs of the American Mathematical Society is published bimonthly (each volume consisting usually of more than one number) by the American Mathematical Society at 201 Charles Street, Providence, RI 02904-2294, USA. Periodicals postage paid at Providence, RI. Postmaster: Send address changes to Memoirs, American Mathematical Society, 201 Charles Street, Providence, RI 02904-2294, USA.

This publication is indexed in *Science Citation Index*®, *SciSearch*®, *Research Alert*®, *CompuMath Citation Index*®, *Current Contents*®/*Physical, Chemical & Earth Sciences.*
Printed in the United States of America.

♾ The paper used in this book is acid-free and falls within the guidelines established to ensure permanence and durability.
Visit the AMS home page at `http://www.ams.org/`

10 9 8 7 6 5 4 3 2 1 11 10 09 08 07 06

Contents

Abstract

We define a new notion of entropy for operators on Fock spaces and positive multi-Toeplitz kernels on free semigroups. This is studied in connection with factorization theorems for (eg. multi-Toeplitz, multi-analytic, etc.) operators on Fock spaces. These results lead to entropy inequalities and entropy formulas for positive multi-Toeplitz kernels on free semigroups (resp. multi-analytic operators) and consequences concerning the extreme points of the unit ball of the noncommutative analytic Toeplitz algebra F_n^∞.

We obtain several geometric characterizations of the central intertwining lifting, a maximal principle, and a permanence principle for the noncommutative commutant lifting theorem. Under certain natural conditions, we find explicit forms for the maximal entropy solution of this multivariable commutant lifting theorem.

All these results are used to solve maximal entropy interpolation problems in several variables. We obtain explicit forms for the maximal entropy solution (as well as its entropy) of the Sarason, Carathéodory-Schur, and Nevanlinna-Pick type interpolation problems for the noncommutative (resp. commutative) analytic Toeplitz algebra F_n^∞ (resp. W_n^∞) and their tensor products with $B(\mathcal{H}, \mathcal{K})$. In particular, we provide explicit forms for the maximal entropy solutions of several interpolation problems on the unit ball of $\mathbb{C}^n$.

Received by the editor May 20, 2003.

2000 *Mathematics Subject Classification.* Primary: 47A57; 47A13; Secondary: 47A56; 47A20; 47B35.

Key words and phrases. Entropy; Interpolation; Factorization; Commutant lifting; Row contraction; Isometric dilation; Fock space; Free semigroup; Positive definite kernels; Multivariable operator theory; Analytic operator; Toeplitz operator.

Research supported in part by an NSF grant.

Introduction

Let H_n be an n-dimensional complex Hilbert space with orthonormal basis e_1, $e_2, \ldots, e_n$, where $n \in \{1, 2, \ldots\}$ or $n = \infty$. We consider the full Fock space of H_n defined by

$$F^2(H_n) := \bigoplus_{k \geq 0} H_n^{\otimes k},$$

where $H_n^{\otimes 0} := \mathbb{C}1$ and $H_n^{\otimes k}$ is the (Hilbert) tensor product of k copies of H_n. Define the left creation operators $S_i : F^2(H_n) \to F^2(H_n)$, $i = 1, \ldots, n$, by

$$S_i \psi := e_i \otimes \psi, \ \psi \in F^2(H_n).$$

The noncommutative analytic Toeplitz algebra F_n^∞ and its norm closed version (the noncommutative disc algebra $\mathcal{A}_n$) were introduced by the author [**42**] in connection with a multivariable noncommutative von Neumann inequality. F_n^∞ is the algebra of left multipliers of the full Fock space $F^2(H_n)$ and can be identified with the weakly closed (or w^*-closed) algebra generated by the left creation operators $S_1, \ldots, S_n$ on the full Fock space $F^2(H_n)$, and the identity. The noncommutative disc algebra $\mathcal{A}_n$ is the norm closed algebra generated by the left creation operators $S_1, \ldots, S_n$ and the identity. When $n = 1$, F_1^∞ can be identified with $H^\infty(\mathbb{D})$, the algebra of bounded analytic functions on the open unit disc. The algebra F_n^∞ can be viewed as a multivariable noncommutative analogue of $H^\infty(\mathbb{D})$. We should add that the algebra F_n^∞ shares many properties with $H^\infty(\mathbb{D})$. There are many analogies with the invariant subspaces of the unilateral shift on $H^2(\mathbb{D})$, inner-outer factorizations, analytic operators, Toeplitz operators, $H^\infty(\mathbb{D})$–functional calculus, bounded (resp. spectral) interpolation, etc. The noncommutative analytic Toeplitz algebra F_n^∞ has been studied in several papers [**39**], [**40**], [**43**], [**44**], [**45**], [**47**], [**2**], and recently in [**14**], [**15**], [**16**], [**4**], [**51**], [**13**], [**36**], and [**56**].

We established a strong connection between the algebra F_n^∞ and the function theory on the open unit ball

$$\mathbb{B}_n := \{(\lambda_1, \ldots, \lambda_n) \in \mathbb{C}^n : \ |\lambda_1|^2 + \cdots + |\lambda_n|^2 < 1\},$$

through the noncommutative von Neumann inequality [**42**] (see also [**43**], [**45**], [**47**], and [**50**]). In particular, we proved that there is a completely contractive homomorphism $\Phi : F_n^\infty \to H^\infty(\mathbb{B}_n)$ defined by

$$[\Phi(f(S_1, \ldots, S_n))](\lambda_1, \ldots, \lambda_n) = f(\lambda_1, \ldots, \lambda_n)$$

for any $f(S_1, \ldots, S_n) \in F_n^\infty$ and $(\lambda_1, \ldots, \lambda_n) \in \mathbb{B}_n$. A characterization of the analytic functions in the range of the map Φ was obtained in [**4**], and independently in [**16**]. Moreover, it was proved that the quotient $F_n^\infty / \ker \Phi$ is an operator algebra which can be identified with $W_n^\infty := P_{F_s^2(H_n)} F_n^\infty |_{F_s^2(H_n)}$, the compression of F_n^∞ to the symmetric Fock space $F_s^2(H_n) \subset F^2(H_n)$. In [**47**], [**6**], [**7**], [**4**], [**3**], [**16**], [**50**],

[**51**] and [**36**], a good case is made that the appropriate multivariable commutative analogue of $H^\infty(\mathbb{D})$ is the algebra W_n^∞, which was also proved to be the w^*-closed algebra generated by the operators $B_i := P_{F_s^2(H_n)} S_i|_{F_s^2(H_n)}$, $i = 1, \ldots, n$, and the identity.

Moreover, Arveson showed in [**6**] that W_n^∞ can be seen as the algebra of all analytic multipliers of $F_s^2(H_n)$, when $F_s^2(H_n)$ is identified with a class of analytic functions in $\mathbb{B}_n$. More precisely, the range of the homomorphism Φ is the multiplier algebra of the reproducing kernel Hilbert space with reproducing kernel $K_n : \mathbb{B}_n \times \mathbb{B}_n \to \mathbb{C}$ defined by

$$K_n(z, w) := \frac{1}{1 - \langle z, w\rangle_{\mathbb{C}^n}}, \qquad z, w \in \mathbb{B}_n.$$

Interpolation problems for the noncommutative analytic Toeplitz algebra F_n^∞ were first considered by the author in [**39**] and [**44**], where we obtained Sarason [**60**] and Carathéodory [**12**] type interpolation theorems, respectively. In 1997, Arias and the author [**4**] (see also [**3**] and [**46**]) extended Sarason's result [**60**] and obtained a distance formula to an arbitrary WOT-closed ideal in F_n^∞ as well as a Nevanlinna-Pick type interpolation theorem (see [**35**]) for the noncommutative analytic Toeplitz algebra F_n^∞. Using different methods, Davidson and Pitts proved these results independently in [**16**]. Let us mention that, recently, interpolation problems for F_n^∞ (resp. W_n^∞) and interpolation problems on the unit ball $\mathbb{B}_n$ were also considered in [**1**], [**49**], [**50**], [**52**], [**8**], [**53**], [**54**], [**55**], [**9**], and [**20**] .

Due to the connection between the algebras F_n^∞, W_n^∞, and $H^\infty(\mathbb{B}_n)$, the interpolation problems for the noncommutative analytic Toeplitz algebra F_n^∞ have appropriate versions for the commutative Toeplitz algebra W_n^∞ and the Hardy space $H^\infty(\mathbb{B}_n)$ (resp. some classes of bounded analytic functions in $\mathbb{B}_n$). This claim is supported by the following papers: [**3**], [**4**], [**46**], [**49**], [**50**], [**52**], [**53**], [**54**], and [**55**].

Let $\Theta : F^2(H_n) \otimes \mathcal{H} \to F^2(H_n) \otimes \mathcal{K}$ be a multi-analytic operator, i.e.,

$$\Theta(S_i \otimes I_{\mathcal{H}}) = (S_i \otimes I_{\mathcal{K}})\Theta, \qquad i = 1, \ldots, n.$$

We recall that (see [**44**], [**53**]) that $\Theta \in R_n^\infty \bar{\otimes} B(\mathcal{H}, \mathcal{K})$, where R_n^∞ is the weakly closed algebra generated by the right creation operators on the full Fock space, and the identity. Moreover, we have $R_n^\infty = U^* F_n^\infty U$, where U is a unitary operator (see Section 1.1). If $\|\Theta\| \leq 1$ and $\dim \mathcal{H} < \infty$, then we define the prediction entropy of the multi-analytic operator Θ by setting

$$E(\Theta) := \ln \det \Delta(\Theta),$$

where

$$\langle \Delta(\Theta)x, x\rangle := \inf\{\langle (I - \Theta^*\Theta)(x - p), x - p\rangle : \; p \in F^2(H_n) \otimes \mathcal{H}, \; p(0) = 0\}$$

for any $x \in \mathcal{H}$. It turns out (using Szegö's theorem) that in the particular case when $n = 1$ and $\mathcal{H} = \mathcal{K} = \mathbb{C}$ we have

$$E(f) = \frac{1}{2\pi} \int_{-\pi}^{\pi} \ln(1 - |f(e^{it})|^2)\, dt,$$

which is the classical definition for the entropy of $f \in H^\infty(\mathbb{D})$ with $\|f\| \leq 1$.

There is an extensive literature (see [**11**], [**5**], [**17**], [**18**], [**21**], [**19**], [**22**], [**23**], [**58**], etc.) concerning the maximal entropy solutions for the classical interpolation problems of Carathéodory-Schur ([**12**], [**61**]), Nevanlinna-Pick ([**35**]), Nehari ([**34**]), Sarason ([**60**]), and the more general commutant lifting theorem of Sz.-Nagy and Foiaş ([**65**]). Maximal entropy solutions have also played an important role in control theory (see [**26**], [**27**], [**25**], etc.).

The main goal of this paper is to find the maximal entropy solution in the noncommutative commutant lifting theorem (see [**38**], [**41**]) and to use it in order to get the maximal entropy solutions for the Sarason, Carathéodory-Schur, and Nevanlinna-Pick type interpolation problems for the noncommutative (resp. commutative) analytic Toeplitz algebra F_n^∞ (resp. W_n^∞). Moreover, under certain natural conditions, we obtain explicit forms for the maximal entropy solutions and their entropy. In particular, we solve maximal entropy interpolation problems on the unit ball of $\mathbb{C}^n$.

Our investigation is based on multivariable noncommutative dilation theory (see [**10**], [**24**], [**37**], [**38**], [**39**], [**41**], [**48**]), harmonic analysis on Fock spaces (see [**39**], [**40**], [**41**], [**44**], [**51**], [**2**], [**14**], and [**15**]), the results of Foiaş, Frazho, and Gohberg [**22**] (see also [**21**], [**23**], and [**66**]) on maximal entropy interpolation (case $n = 1$), and the classical results on interpolation and commutant lifting theorem (see [**12**], [**61**], [**35**], [**34**], [**60**], [**65**], [**66**], [**21**], [**23**], etc.)

The paper is organized in three chapters. In the first chapter, we define a new notion of entropy for operators on Fock spaces and positive definite multi-Toeplitz kernels on free semigroups. This is studied in connection with factorization theorems for (multi-Toeplitz, multi-analytic, etc.) operators on Fock spaces. These results lead to entropy inequalities and entropy formulas for positive definite multi-Toeplitz kernels on free semigroups (resp. multi-Toeplitz operators) and consequences concerning the extreme points of the unit ball of the noncommutative analytic Toeplitz algebra F_n^∞.

More precisely, in Section 1.1, we prove the existence of maximal outer factors for arbitrary positive multi-Toeplitz operators on Fock spaces. A connection between the Szegö infimum and maximal outer factors of multi-Toeplitz operators via the prediction-error operators is considered. These results extend some classical results (see [**62**], [**63**], [**64**], [**28**], [**29**], [**30**], [**32**], [**33**], [**66**], [**67**], [**68**], and [**69**]) as well as some extensions to Fock spaces from ([**44**] and [**48**]). We introduce the notion of prediction entropy for positive multi-Toeplitz operators $T \in B(F^2(H_n) \otimes \mathcal{E})$, where $\mathcal{E}$ is a Hilbert space with $\dim \mathcal{E} < \infty$. In particular, we prove that the entropy of T satisfies the equation

$$e(T) = \ln \det[\varphi(0)^* \varphi(0)],$$

where M_φ is the maximal outer factor of T. Finally, we provide an explicit form for the square outer spectral factor M_φ corresponding to a strictly positive multi-Toeplitz operator T, i.e., $T = M_\varphi^* M_\varphi$.

In the next section, we provide factorization results for bounded linear operators in $B(\mathcal{H}_1, F^2(H_n) \otimes \mathcal{H}_2)$, where $\mathcal{H}_1$ is a finite dimensional Hilbert space, and for multi-analytic operators, generalizing classical results from [**29**], [**30**], [**59**], and [**67**], as well as some extensions to Fock spaces (see [**40**], [**44**], and [**48**]). A noncommutative multivariable analogue of Robinson's minimum energy delay principle (see [**57**] and [**22**]) for outer operators on Fock spaces is obtained.

These results are used to prove entropy inequalities for positive definite multi-Toeplitz kernels on the free semigroup with n generators $\mathbb{F}_n^+$ and multi-Toeplitz operators on Fock spaces. We extend the classical result (see [**31**]) which stated for the Hardy space $H^2(\mathbb{D})$ says that if $f \in H^2(\mathbb{D})$, then $\ln|f(e^{it})|$ is integrable and

$$\frac{1}{2\pi}\int_{-\pi}^{\pi} \ln|f(e^{it})|\,dt \geq \ln|f(0)|. \tag{0.1}$$

Next, we give a characterization for the outer operators in $B(\mathcal{E}, F^2(H_n)\otimes\mathcal{E})$ if $\dim\mathcal{E} < \infty$. In particular, we find a noncommutative multivariable analogue of the classical result saying that a function $f \in H^2(\mathbb{D})$ is outer if and only if $f(0) \neq 0$ and the equality holds in (0.1).

It is well-known [**31**] that a function $f \in H^\infty(\mathbb{D})$ is an extreme point of the unit ball of $H^\infty(\mathbb{D})$ if and only if

$$\frac{1}{2\pi}\int_{-\pi}^{\pi} \ln(1-|f(e^{it})|^2)\,dt = -\infty.$$

In Section 1.4, we prove some results concerning the extreme points of the unit ball of F_n^∞. In particular, we show that if $\varphi \in F_n^\infty$, $\|\varphi\| \leq 1$, and the entropy $E(\varphi) = -\infty$, then φ is an extreme point of the unit ball of F_n^∞.

In the second chapter of this paper, we obtain several geometric characterizations for the multivariable central intertwining lifting, a maximum principle, and a permanence principle for the noncommutative commutant lifting theorem. Under certain natural conditions, we find explicit forms for the maximal entropy solution of this multivariable commutant lifting theorem, and concrete formulas for its entropy.

Let us recall the noncommutative commutant lifting theorem for row contractions [**38**], [**41**] (see [**65**] for the classical case $n = 1$). Let $\mathcal{T} := [T_1 \cdots T_n]$, $T_i \in B(\mathcal{H})$, and $\mathcal{T}' := [T_1' \cdots T_n']$, $T_i' \in B(\mathcal{H}')$, be row contractions and let $\mathcal{V} := [V_1 \cdots V_n]$, $V_i \in B(\mathcal{K})$, and $\mathcal{V}' := [V_1' \cdots V_n']$, $V_i' \in B(\mathcal{K}')$, be the corresponding minimal isometric dilations. The noncommutative commutant lifting theorem states that if $A \in B(\mathcal{H}, \mathcal{H}')$ is an operator satisfying

$$AT_i = T_i'A, \qquad i = 1, \ldots, n,$$

then there exists an operator $B \in B(\mathcal{K}, \mathcal{K}')$ with the following properties:

(i) $BV_i = V_i'B$ for any $i = 1, \ldots, n$;
(ii) $B^*|\mathcal{H}' = A^*$;
(iii) $\|B\| = \|A\|$.

As in the classical case, the general setting of the noncommutative commutant lifting theorem can be reduced to the case when $\mathcal{T} := [T_1 \cdots T_n]$ is a row isometry (see [**55**]).

In Section 2.1, we present some results concerning the geometric structure of the intertwining liftings in the noncommutative commutant lifting theorem. It is shown that there is a one-to-one correspondence between the set of all multivariable intertwining liftings B with tolerance $t > 0$ (i.e., $\|B\| \leq t$) and certain families of contractions $\{C_k\}_{k=1}^\infty$ and $\{\Lambda_\alpha\}_{\alpha\in\mathbb{F}_n^+}$. Moreover, we prove that the intertwining lifting corresponding to the parameters $C_k = 0$, $k = 1, 2, \ldots$, (resp. $\Lambda_\alpha = 0$, $\alpha \in \mathbb{F}_n^+$) coincides with the central intertwining lifting with tolerance t, for which we have an explicit form.

The geometric structure of the central intertwining lifting will play a very important role in our investigation. We prove a maximum principle for the noncommutative commutant lifting theorem, which also provides a new characterization for the central intertwining lifting. This result is used to prove a permanence principle for the central intertwining lifting, which generalizes the permanence principle for the Carathéodory interpolation problem (see **[19]** and **[22]**) (case $n = 1$). Applications of this principle will be considered in the last chapter.

The next step is to obtain explicit formulas for the quasi outer spectral factor for the defect operator $t^2 I - B_c^* B_c$ of the central intertwining lifting B_c with tolerance $t > 0$. This leads, in the next section, to concrete formulas for the entropy of B_c as well as to a maximum principle for the noncommutative commutant lifting theorem with respect to non-minimal isometric liftings.

In Section 2.6, we present one of the main results of this paper. Using the maximum principle, we prove that the central intertwining lifting B_c is the maximal entropy solution for the noncommutative commutant lifting theorem, when $\mathcal{T} := [S_1 \otimes I_{\mathcal{E}} \ \cdots \ S_n \otimes I_{\mathcal{E}}]$ with $\dim \mathcal{E} < \infty$. Based on several results of this paper, we are led to concrete formulas for the entropy of B_c and, under a certain condition of stability, to a maximum principle and a characterization (in terms of entropy) of the central intertwining lifting $\tilde{B}_c$ with respect to non-minimal isometric liftings.

The results of the first two chapters are used to solve maximal entropy interpolation problems in several variables, in the last chapter of this paper. We obtain explicit forms for the maximal entropy solution (as well as its entropy) of the Sarason **[60]**, Carathéodory-Schur **[12]**, **[61]**, and Nevanlinna-Pick **[35]** type interpolation problems for the noncommutative (resp. commutative) analytic Toeplitz algebra F_n^∞ (resp. W_n^∞) and their tensor products with $B(\mathcal{H}, \mathcal{K})$, the set of all bounded linear operators acting on Hilbert spaces. Moreover, under certain conditions, we also find explicit forms for the corresponding classical optimization problems, in our multivariable noncommutative (resp. commutative) setting.

In particular, we provide explicit forms for the maximal entropy solutions of several interpolation problems on the unit ball of $\mathbb{C}^n$. Finnaly, we apply our permanence principle to the Nevanlinna-Pick interpolation problem on the unit ball.

CHAPTER 1

Operators on Fock Spaces and their Entropy

A new notion of entropy for operators on Fock spaces and positive multi-Toeplitz kernels on free semigroups is defined and studied in connection with factorization theorems for (multi-Toeplitz, multi-analytic, etc.) operators on Fock spaces. These results lead to entropy inequalities and entropy formulas for positive multi-Toeplitz kernels on free semigroups (resp. multi-Toeplitz operators) and consequences concerning the extreme points of the unit ball of the noncommutative analytic Toeplitz algebra F_n^∞.

1.1. Entropy and spectral factorization for multi-Toeplitz operators

In this section, we define the notion of prediction entropy for positive multi-Toeplitz operators $T \in B(F^2(H_n) \otimes \mathcal{E})$ with $\dim \mathcal{E} < \infty$. We prove that there is a multi-analytic operator $M_\varphi \in R_n^\infty \bar{\otimes} B(\mathcal{E})$ such that $M_\varphi^* M_\varphi \leq T$ if and only if the entropy of T satisfies $e(T) > -\infty$. This is based on Theorem 1.1 which proves the existence of maximal outer factors for arbitrary positive multi-Toeplitz operators on Fock spaces. Moreover, we prove a Szegö type infimum theorem for arbitrary positive multi-Toeplitz operators (see Theorem 1.3), and provide an explicit form for the square outer spectral factor M_φ corresponding to a strictly positive multi-Toeplitz operator T, i.e., $M_\varphi^* M_\varphi = T$ (see Theorem 1.5). All these results are needed in the next sections.

Let $\mathbb{F}_n^+$ be the unital free semigroup on n generators $g_1, \ldots, g_n$, and the identity g_0. The length of $\alpha \in \mathbb{F}_n^+$ is defined by $|\alpha| := k$, if $\alpha = g_{i_1} g_{i_2} \cdots g_{i_k}$, and $|\alpha| := 0$, if $\alpha = g_0$. We also define $e_\alpha := e_{i_1} \otimes e_{i_2} \otimes \cdots \otimes e_{i_k}$ and $e_{g_0} = 1$. It is clear that $\{e_\alpha : \alpha \in \mathbb{F}_n^+\}$ is an orthonormal basis of the full Fock space $F^2(H_n)$.

Let $\mathcal{E}$ be a Hilbert space and let T be a positive multi-Toeplitz operator on $F^2(H_n) \otimes \mathcal{E}$, i.e.,

$$(S_i \otimes I_\mathcal{E})^* T (S_j \otimes I_\mathcal{E}) = \delta_{ij} I$$

for any $i, j = 1, \ldots, n$. Define the positive operator $\Delta_T : \mathcal{E} \to \mathcal{E}$ by setting

$$\langle \Delta_T x, x \rangle := \inf\{\langle T(x-p), x-p \rangle : \ p \in \mathcal{P}(\mathcal{E}),\ p(0) = 0\} \tag{1.1}$$

for any $x \in \mathcal{E}$, where $\mathcal{P}(\mathcal{E})$ denotes the set of polynomials $p = \sum_{|\alpha| \leq m} e_\alpha \otimes h_\alpha$, $m = 0, 1, \ldots$, in $F^2(H_n) \otimes \mathcal{E}$ and $p(0) := h_{g_0}$. We remark that if T is a normalized positive multi-Toeplitz operator, i.e., $P_\mathcal{E} T|\mathcal{E} = I_\mathcal{E}$, then Δ_T coincides with the prediction-error operator of the stationary process determined by T (see [**48**]).

When $\mathcal{E}$ is finite dimensional, we define the prediction entropy of the positive multi-Toeplitz operator T by

$$e(T) := \ln \det \Delta_T. \tag{1.2}$$

Note that the prediction entropy is different from the entropy defined in [**48**].

We need to recall from [**39**], [**40**], [**42**], [**43**], and [**44**] a few facts concerning multi-analytic operators on Fock spaces. We say that $M \in B(F^2(H_n)\otimes\mathcal{K}, F^2(H_n)\otimes\mathcal{K}')$ is multi-analytic if

$$M(S_i \otimes I_{\mathcal{K}}) = (S_i \otimes I_{\mathcal{K}'})M \quad \text{for any } \ i = 1, \dots, n. \tag{1.3}$$

Note that M is uniquely determined by the operator $\theta : \mathcal{K} \to F^2(H_n) \otimes \mathcal{K}'$, which is defined by $\theta k := M(1 \otimes k)$, $k \in \mathcal{K}$, and is called the symbol of M. We denote $M = M_\theta$. Moreover, M_θ is uniquely determined by the "coefficients" $\theta_\alpha \in B(\mathcal{K}, \mathcal{K}')$, which are given by

$$\langle \theta_{\tilde\alpha} k, k'\rangle := \langle \theta k, e_\alpha \otimes k'\rangle = \langle M_\theta(1\otimes k), e_\alpha \otimes k'\rangle\,, \quad k \in \mathcal{K},\ k' \in \mathcal{K}',\ \alpha \in \mathbb{F}_n^+,$$

where $\tilde\alpha$ is the reverse of α, i.e., $\tilde\alpha = g_{i_k}\cdots g_{i_1}$ if $\alpha = g_{i_1}\cdots g_{i_k}$. Note that

$$\sum_{\alpha\in\mathbb{F}_n^+} \theta_\alpha^*\theta_\alpha \le \|M_\theta\|^2 I_{\mathcal{K}}.$$

If $T_1, \dots, T_n \in B(\mathcal{H})$ (the algebra of all bounded linear operators on the Hilbert space $\mathcal{H}$), define $T_\alpha := T_{i_1}T_{i_2}\cdots T_{i_k}$, if $\alpha = g_{i_1}g_{i_2}\cdots g_{i_k}$ and $T_{g_0} := I_{\mathcal{H}}$. We can associate with M_θ a unique formal Fourier expansion

$$M_\theta \sim \sum_{\alpha\in\mathbb{F}_n^+} R_\alpha \otimes \theta_\alpha, \tag{1.4}$$

where $R_i := U^*S_iU$, $i = 1, \dots, n$, are the right creation operators on $F^2(H_n)$ and U is the (flipping) unitary operator on $F^2(H_n)$ mapping $e_{i_1}\otimes e_{i_2}\otimes\cdots\otimes e_{i_k}$ into $e_{i_k}\otimes\cdots\otimes e_{i_2}\otimes e_{i_1}$. Since M_θ acts like its Fourier representation on "polynomials", we will identify them for simplicity. The set of multi-analytic operators in $B(F^2(H_n)\otimes\mathcal{K}, F^2(H_n)\otimes\mathcal{K}')$ coincides with $R_n^\infty\bar\otimes B(\mathcal{K},\mathcal{K}')$, where $R_n^\infty = U^*F_n^\infty U$ (see [**44**] and [**53**]).

A multi-analytic operator M_θ (resp. its symbol θ) is called inner if M_θ is an isometry. We call M_θ (resp. its symbol θ) outer if

$$\bigvee\{(S_\alpha \otimes I_{\mathcal{K}'})\theta k :\ k \in \mathcal{K}, \alpha \in \mathbb{F}_n^+\} = F^2(H_n)\otimes\mathcal{K}'.$$

We say that a positive multi-Toeplitz operator $T \in B(F^2(H_n)\otimes\mathcal{E})$ has a maximal outer factor if there is a Hilbert space $\mathcal{G}$ and an outer multi-analytic operator $M_\varphi \in R_n^\infty\bar\otimes B(\mathcal{E},\mathcal{G})$ with the properties:

(i) $M_\varphi^* M_\varphi \le T$;

(ii) If $\mathcal{N}$ is a Hilbert space and $M_\theta \in R_n^\infty\bar\otimes B(\mathcal{E},\mathcal{N})$ satisfies

$$M_\theta^* M_\theta \le T,$$

then $M_\varphi^* M_\varphi \ge M_\theta^* M_\theta$.

In what follows we prove the existence of maximal outer factors for arbitrary positive multi-Toeplitz operators on Fock spaces.

Theorem 1.1. *If $\mathcal{E}$ is an arbitrary Hilbert space and $T \in B(F^2(H_n)\otimes\mathcal{E})$ is a positive multi-Toeplitz operator, then T has a maximal outer factor $M_\varphi \in R_n^\infty\bar\otimes B(\mathcal{E},\mathcal{G})$ such that*

$$M_\varphi^* M_\varphi \le T. \tag{1.5}$$

Moreover, the maximal outer factor is uniquely determined up to a unitary diagonal multi-analytic operator.

PROOF. Since T is a positive multi-Toeplitz operator, we have

$$\begin{aligned}\left\| T^{1/2}\left(\sum_{i=1}^{n}(S_i\otimes I_{\mathcal{E}})h_i\right)\right\|^2 &= \sum_{i,j=1}^{n}\langle (S_j^*\otimes I_{\mathcal{E}})T(S_i\otimes I_{\mathcal{E}})h_i, h_j\rangle\\ &= \sum_{i,j=1}^{n}\langle \delta_{ij}Th_i, h_j\rangle = \sum_{i=1}^{n}\|T^{1/2}h_i\|^2\end{aligned}$$

for any $h_i \in F^2(H_n)\otimes\mathcal{E}$, $i=1,\dots,n$. Let $\mathcal{Y} := \overline{T^{1/2}(F^2(H_n)\otimes\mathcal{E})}$ and note that there are unique isometries $V_i\in B(\mathcal{Y})$, $i=1,\dots,n$, such that

$$V_iT^{1/2} = T^{1/2}(S_i\otimes I_{\mathcal{E}}),\quad i=1,\dots,n. \tag{1.6}$$

The above calculations show that

$$\left\|\sum_{i=1}^{n} V_iT^{1/2}h_i\right\|^2 = \sum_{i=1}^{n}\|T^{1/2}h_i\|^2$$

for any $h_i\in\mathcal{Y}$. Hence, $V_1,\dots,V_n$ are isometries with orthogonal ranges. According to the Wold type decomposition for isometries with orthogonal ranges (see [**38**]), we have $V_i = U_i\oplus W_i$, $i=1,\dots,n$, with respect to the orthogonal decomposition $\mathcal{Y}=\mathcal{Y}_0\oplus\mathcal{Y}_1$, where $\mathcal{Y}_0 := \oplus_{\alpha\in\mathbb{F}_n^+}V_\alpha\mathcal{M}_T$ and

$$\mathcal{M}_T = \mathcal{Y}\ominus\bigvee_{i=1}^{n} T^{1/2}(S_i\otimes I_{\mathcal{E}})(F^2(H_n)\otimes\mathcal{E}). \tag{1.7}$$

The subspaces $\mathcal{Y}_0$ and $\mathcal{Y}_1$ are reducing for each operator V_i, $i=1,\dots,n$. Moreover,

$$\sum_{i=1}^{n} W_iW_i^* = I_{\mathcal{Y}_1}$$

and $\{U_i\}_{i=1}^n$ is unitarily equivalent to the orthogonal shift $\{S_i\otimes I_{\mathcal{M}_T}\}_{i=1}^n$. Hence, we have

$$\Phi U_i = (S_i\otimes I_{\mathcal{M}_T})\Phi,\quad i=1,\dots,n,$$

where $\Phi:\mathcal{Y}_0\to F^2(H_n)\otimes\mathcal{M}_T$ is the Fourier transform defined by

$$\Phi(V_\alpha\ell_\alpha) = e_\alpha\otimes\ell_\alpha,\quad \ell_\alpha\in\mathcal{M}_T,\ \alpha\in\mathbb{F}_n^+. \tag{1.8}$$

Since $\mathcal{Y}_0$ reduces each operator V_i, $i=1,\dots,n$, it follows that

$$U_iP_{\mathcal{Y}_0} = V_iP_{\mathcal{Y}_0} = P_{\mathcal{Y}_0}V_i,\quad i=1,\dots,n.$$

Hence, we infer that

$$\begin{aligned}\Phi P_{\mathcal{Y}_0}T^{1/2}(S_i\otimes I_{\mathcal{E}}) &= \Phi P_{\mathcal{Y}_0}V_iT^{1/2} = \Phi U_iP_{\mathcal{Y}_0}T^{1/2}\\ &= (S_i\otimes I_{\mathcal{M}_T})\Phi P_{\mathcal{Y}_0}T^{1/2}\end{aligned}$$

for any $i=1,\dots,n$. This shows that the operator $\Phi P_{\mathcal{Y}_0}T^{1/2}$ is multi-analytic and, according to [**44**], there exists $M_\varphi\in R_n^\infty\bar{\otimes}B(\mathcal{E},\mathcal{M}_T)$ such that

$$\Phi P_{\mathcal{Y}_0}T^{1/2} = M_\varphi. \tag{1.9}$$

Since the operator $P_{\mathcal{Y}_0}T^{1/2}$ has dense range in $\mathcal{Y}_0$, it follows that M_φ is outer. Finally, relation (1.9) impies $M_\varphi^*M_\varphi\le T$.

To prove the maximality property, let $\mathcal{N}$ be a Hilbert space and let $M_\theta \in R_n^\infty \bar{\otimes} B(\mathcal{E}, \mathcal{N})$ be such that $M_\theta^* M_\theta \leq T$. Define $X : \mathcal{Y} \to F^2(H_n) \otimes \mathcal{E}$ by setting

$$X\left(\sum_{|\sigma|\leq m} V_\sigma T^{1/2} y_\sigma\right) := \sum_{|\sigma|\leq m} (S_\sigma \otimes I_{\mathcal{N}})\theta y_\sigma, \quad y_\sigma \in \mathcal{E}.$$

Note that, for any $y_\sigma \in \mathcal{E}$, we have

$$\begin{aligned}
\left\| X\left(\sum_{|\sigma|\leq m} V_\sigma T^{1/2} y_\sigma\right)\right\|^2 &= \left\langle M_\theta^* M_\theta \left(\sum_{|\sigma|\leq m} e_\sigma \otimes y_\sigma\right), \sum_{|\sigma|\leq m} e_\sigma \otimes y_\sigma \right\rangle \\
&\leq \left\langle T\left(\sum_{|\sigma|\leq m} e_\sigma \otimes y_\sigma\right), \sum_{|\sigma|\leq m} e_\sigma \otimes y_\sigma \right\rangle \\
&= \left\| \sum_{|\sigma|\leq m} V_\sigma T^{1/2} y_\sigma \right\|^2.
\end{aligned}$$

This shows that X extends to a contraction from $\mathcal{Y}$ to $F^2(H_n) \otimes \mathcal{N}$. One can easily check that $XV_i = (S_i \otimes I_{\mathcal{N}})X$, $i = 1, \ldots, n$. Hence, and using the Wold decomposition from [**38**], we deduce

$$\begin{aligned}
X\mathcal{Y}_1 \subseteq \bigcap_{k=0}^{\infty} X\left(\bigoplus_{|\alpha|=k} V_\alpha \mathcal{Y}\right) &= \bigcap_{k=0}^{\infty} \bigoplus_{|\alpha|=k} (S_\alpha \otimes I_{\mathcal{E}})X\mathcal{Y} \\
&\subseteq \bigcap_{k=0}^{\infty} \bigoplus_{|\alpha|=k} (S_\alpha \otimes I_{\mathcal{E}})(F^2(H_n) \otimes \mathcal{E}) = \{0\}.
\end{aligned}$$

Therefore, $X|\mathcal{Y}_1 = 0$. Now, taking into account the above considerations, we have

$$\begin{aligned}
\left\langle M_\theta^* M_\theta \left(\sum_{|\sigma|\leq m} e_\sigma \otimes y_\sigma\right), \sum_{|\sigma|\leq m} e_\sigma \otimes y_\sigma \right\rangle &= \left\| X\left(\sum_{|\sigma|\leq m} P_{\mathcal{Y}_0} V_\sigma T^{1/2} y_\sigma\right)\right\|^2 \\
&\leq \left\| \sum_{|\sigma|\leq m} \Phi P_{\mathcal{Y}_0} V_\sigma T^{1/2} y_\sigma \right\|^2 \\
&= \left\| \sum_{|\sigma|\leq m} \Phi P_{\mathcal{Y}_0} T^{1/2} (S_\sigma \otimes I_{\mathcal{E}})(1 \otimes y_\sigma) \right\|^2 \\
&= \left\| \sum_{|\sigma|\leq m} M_\varphi (S_\sigma \otimes I_{\mathcal{E}})(1 \otimes y_\sigma) \right\|^2 \\
&= \left\langle M_\varphi^* M_\varphi \left(\sum_{|\sigma|\leq m} e_\sigma \otimes y_\sigma\right), \sum_{|\sigma|\leq m} e_\sigma \otimes y_\sigma \right\rangle.
\end{aligned}$$

Therefore, $M_\theta^* M_\theta \leq M_\varphi^* M_\varphi$. The proof is complete. $\square$

Let us remark that the equality holds in (1.5) if and only if $\mathcal{Y}_0 = \mathcal{Y}$. Another characterization of this fact can be found in [**40**]. On the other hand, we have a concrete form for the maximal outer factor of T, that is, $M_\varphi = \Phi P_{\mathcal{Y}_0} T^{1/2}$, where the Fourier transform Φ and the subspace $\mathcal{Y}_0$ are defined in the proof of Theorem 1.1.

COROLLARY 1.2. *Let* $T \in B(F^2(H_n) \otimes \mathcal{E})$ *be a positive multi-Toeplitz operator and assume that* $\dim \mathcal{E} < \infty$. *Then the entropy* $e(T) > -\infty$ *if and only if there exists an outer multi-analytic operator* $M_\varphi \in R_n^\infty \bar{\otimes} B(\mathcal{E})$ *such that*

$$M_\varphi^* M_\varphi \leq T. \tag{1.10}$$

PROOF. Assume now that $T = M_\varphi^* M_\varphi$ with $M_\varphi \in R_n^\infty \otimes B(\mathcal{E})$ an outer operator. Then there exists a unitary operator $Z : \mathcal{Y} \to F^2(H_n) \otimes \mathcal{E}$ satisfying

$$ZT^{1/2} = M_\varphi.$$

Since M_φ commutes with each operator $S_i \otimes I_\mathcal{E}$, $i = 1, \ldots, n$, relation (1.6) implies

$$\begin{aligned} ZV_i T^{1/2} &= ZT^{1/2}(S_i \otimes I_\mathcal{E}) = M_\varphi(S_i \otimes I_\mathcal{E}) \\ &= (S_i \otimes I_\mathcal{E}) M_\varphi = (S_i \otimes I_\mathcal{E}) Z T^{1/2} \end{aligned} \tag{1.11}$$

for any $i = 1, \ldots, n$. Hence, $ZV_i = (S_i \otimes I_\mathcal{E})Z$ for any $i = 1, \ldots, n$, and consequently $\{V_i\}_{i=1}^n$ is an orthogonal shift having the same multiplicity as $\{S_i \otimes I_\mathcal{E}\}_{i=1}^n$. Therefore, we have $\dim \mathcal{E} = \dim \mathcal{M}_T$, where

$$\mathcal{M}_T = \mathcal{Y} \ominus \bigvee_{i=1}^n T^{1/2}(S_i \otimes I_\mathcal{E})(F^2(H_n) \otimes \mathcal{E}) \tag{1.12}$$

is the wandering subspace of $\{V_i\}_{i=1}^n$ (see [**38**]). Note that, if $\dim \mathcal{E} < \infty$, then we have

$$\begin{aligned} \mathcal{M}_T &= \overline{P_{\mathcal{M}_T} T^{1/2}(F^2(H_n) \otimes \mathcal{E})} \\ &= (P_{\mathcal{M}_T} T^{1/2}\mathcal{E}) \bigvee (\bigvee_{i=1}^n P_{\mathcal{M}_T} T^{1/2}(S_i \otimes \mathcal{E})(F^2(H_n) \otimes \mathcal{E}) \\ &= P_{\mathcal{M}_T} T^{1/2}\mathcal{E}. \end{aligned}$$

Hence, it is clear that the following statements are equivalent:

(i) $e(T) > -\infty$;
(ii) $\det \Delta_T \neq 0$;
(iii) $\dim \mathcal{E} = \dim \mathcal{M}_T$.

Since $e(T) = \ln \det \Delta_T$, we deduce $e(T) > -\infty$. Now, if $T \geq M_\varphi^* M_\varphi$, then we have

$$e(T) \geq e(M_\varphi^* M_\varphi) > -\infty.$$

Conversely, assume that the entropy $e(T) > -\infty$. Then $\dim \mathcal{E} = \dim \mathcal{M}_T$ and the result follows from Theorem 1.1. □

It is well-known that in the classical case ($n = 1$) (see [**23**]), we have equality in (1.10). Whether or not this is true if $n \geq 2$ remains an open problem.

In what follows, we prove a Szegö type infimum theorem for arbitrary positive multi-Toeplitz operators. If M_θ is a multi-analytic operator with Fourier expansion (1.4), then we denote $\theta(0) := \theta_{g_0}$.

THEOREM 1.3. *Let $\mathcal{E}$ be a Hilbert space and let $T \in B(F^2(H_n)\otimes\mathcal{E})$ be a positive multi-Toeplitz operator. Then, for any $h \in \mathcal{E}$,*

$$\inf_{p\in\mathcal{P}(\mathcal{E}),\ p(0)=0} \langle T(h-p), h-p\rangle = \langle \varphi(0)^*\varphi(0)h, h\rangle ,$$

where M_φ is the maximal outer factor of T. In particular, if $\dim \mathcal{E} < \infty$, then the entropy of T satisfies the equality

$$e(T) = \ln\det[\varphi(0)^*\varphi(0)].$$

PROOF. First we prove that

$$\Delta_T = P_{\mathcal{E}} T^{1/2} P_{\mathcal{M}_T} T^{1/2}|\mathcal{E}, \tag{1.13}$$

where the operator Δ_T is defined by relation (1.1), and

$$\mathcal{M}_T := \overline{T^{1/2}(F^2(H_n)\otimes\mathcal{E})} \ominus \bigvee_{i=1}^{n} T^{1/2}(S_i\otimes I_{\mathcal{E}})(F^2(H_n)\otimes\mathcal{E}).$$

Indeed, we have

$$\begin{aligned}\langle \Delta_T h, h\rangle &= \inf\{\|T^{1/2}(h-p)\|^2 : p\in\mathcal{P}(\mathcal{E}),\ p(0)=0\}\\ &= \inf\left\{\|T^{1/2}h - y\|^2 :\ y \in \bigvee_{i=1}^{n} T^{1/2}(S_i\otimes I_{\mathcal{E}})(F^2(H_n)\otimes\mathcal{E})\right\}\\ &= \|P_{\mathcal{M}_T}T^{1/2}h\|^2 = \left\langle (P_{\mathcal{E}}T^{1/2}P_{\mathcal{M}_T}T^{1/2}|\mathcal{E})h, h\right\rangle\end{aligned}$$

for any $h \in \mathcal{E}$. Hence, and using relation (1.9), we obtain

$$\begin{aligned}\langle \Delta_T h, h\rangle &= \|P_{\mathcal{M}_T}T^{1/2}h\|^2\\ &= \|P_{1\otimes\mathcal{M}_T}\Phi P_{\mathcal{Y}_0}T^{1/2}h\|^2 = \|P_{1\otimes\mathcal{M}_T}M_\varphi h\|\\ &= \langle \varphi(0)^*\varphi(0)h, h\rangle ,\end{aligned}$$

for any $h \in \mathcal{E}$, where M_φ is the maximal outer factor of T. The last part of the theorem is now obvious. □

Given a multi-Toeplitz operator $T \in B(F^2(H_n)\otimes\mathcal{H})$, we say that an operator $F \in R_n^\infty\bar{\otimes}B(\mathcal{H},\mathcal{Y})$ is a spectral factor of T if $T = F^*F$. If, in addition, $\mathcal{H} = \mathcal{Y}$ and F is outer, then F is called a square outer spectral factor of T.

In [**40**], we proved that any strictly positive multi-Toeplitz operator T admits a spectral factor. Based on the proof of Theorem 1.1, we can deduce a stronger result, namely that T has a square outer spectral factorization.

COROLLARY 1.4. *If $\mathcal{E}$ is a Hilbert space and $T \in B(F^2(H_n)\otimes\mathcal{E})$ is a strictly positive multi-Toeplitz operator, then there exists an outer multi-analytic operator $M_\varphi \in R_n^\infty\bar{\otimes}B(\mathcal{E})$ such that*

$$M_\varphi^* M_\varphi = T.$$

PROOF. As in the proof of Theorem 1.1, we can define the isometries V_i, $i = 1,\dots,n$, by relation (1.6). Since $T^{1/2}$ is invertible, Theorem 2.1 from [**43**] implies that there exists a unitary operator $U : F^2(H_n)\otimes\mathcal{E} \to \mathcal{Y}$ such that

$$U^*V_iU = S_i\otimes I_{\mathcal{E}}, \quad i = 1,\dots,n.$$

This shows that $\mathcal{Y} = \mathcal{Y}_0$ and relation (1.9) becomes $\Phi T^{1/2} = M_\varphi$, where Φ is the unitary operator defined be (1.8). Hence, $M_\varphi^*M_\varphi = T$ and the proof is complete. □

In the next theorem, we provide an explicit form for the square outer spectral factor corresponding to a strictly positive multi-Toeplitz operator.

THEOREM 1.5. *Let $T \in B(F^2(H_n) \otimes \mathcal{E})$ be a strictly positive multi-Toeplitz operator. Then*

$$T = \Theta^* \Theta,$$

where $\Theta \in R_n^\infty \bar{\otimes} B(\mathcal{E})$ is a square outer spectral factor of T given by

$$\Theta := (I \otimes N) M_\psi^{-1},$$

where M_ψ is an invertible multi-analytic operator with symbol $\psi : \mathcal{E} \to F^2(H_n) \otimes \mathcal{E}$ defined by

$$\psi h := T^{-1}(1 \otimes h), \quad h \in \mathcal{E},$$

and $N := (P_\mathcal{E} T^{-1}|\mathcal{E})^{1/2}$.

PROOF. First we show that $\psi\mathcal{E}$ is cyclic for $\{S_i \otimes I_\mathcal{E}\}_{i=1}^n$ on $F^2(H_n) \otimes \mathcal{E}$. Let $x \in F^2(H_n) \otimes \mathcal{E}$ be such that $x \perp (S_\alpha \otimes I_\mathcal{E}) T^{-1}\mathcal{E}$ for any $\alpha \in \mathbb{F}_n^+$. Since T is invertible, there exists $y \in F^2(H_n) \otimes \mathcal{E}$ such that $Ty = x$. Hence, we have

$$(S_\alpha^* \otimes I_\mathcal{E}) Ty \perp T^{-1}\mathcal{E}, \quad \alpha \in \mathbb{F}_n^+. \tag{1.14}$$

In particular, if $\alpha = e$, then we have $Ty \perp T^{-1}\mathcal{E}$. Since T is positive, this implies $y \perp \mathcal{E}$, and therefore

$$y = \left(\sum_{i=1}^n S_i S_i^* \otimes I_\mathcal{E} \right) y. \tag{1.15}$$

Since T is a multi-Toeplitz operator and using relations (1.15) and (1.14), we infer that

$$T^{-1}\mathcal{E} \perp (S_j^* \otimes I_\mathcal{E}) Ty = (S_j^* \otimes I_\mathcal{E}) T \left(\sum_{i=1}^n S_i S_i^* \otimes I_\mathcal{E} \right) y = T(S_j^* \otimes I_\mathcal{E}) y.$$

Therefore, $(S_j^* \otimes I_\mathcal{E}) y \perp \mathcal{E}$ for any $j = 1, \ldots, n$. Hence,

$$(S_j^* \otimes I_\mathcal{E}) y = \left(\sum_{i=1}^n S_i S_i^* \otimes I_\mathcal{E} \right) (S_j^* \otimes I_\mathcal{E}) y,$$

which together with (1.15) imply

$$y = \left(\sum_{|\alpha|=2} S_\alpha S_\alpha^* \otimes I_\mathcal{E} \right) y.$$

By induction, we can prove that

$$y = \left(\sum_{|\alpha|=k} S_\alpha S_\alpha^* \otimes I_\mathcal{E} \right) y$$

for any $k = 1, 2, \ldots$. This shows that $y \perp (S_\alpha \otimes I_\mathcal{E})(1 \otimes \mathcal{E})$ for any $\alpha \in \mathbb{F}_n^+$. Hence, $y = 0$ and $x = 0$. This proves that M_ψ is an outer operator.

Now, let us show that M_ψ is a bounded multi-analytic operator. Since T is a strictly positive multi-Toeplitz operator, we have

$$\begin{aligned}\langle T(S_\alpha \otimes I_\mathcal{E})\psi h, (S_\alpha \otimes I_\mathcal{E})\psi h\rangle &= \langle T\psi h, \psi h\rangle = \left\langle TT^{-1}h, T^{-1}k\right\rangle \\ &= \left\langle T^{-1}h, k\right\rangle = \left\langle N^2 h, k\right\rangle,\end{aligned}$$

for any $h,k\in\mathcal{E}$. On the other hand, if $\alpha,\beta\in\mathbb{F}_n^+$ and $\beta>\alpha$, i.e., there exists $\tau\in\mathbb{F}_n^+$ such that $\beta=\alpha\tau$, then we have

$$\begin{aligned}\langle T(S_\alpha\otimes I_\mathcal{E})\psi h,(S_\beta\otimes I_\mathcal{E})\psi k\rangle &= \left\langle (S^*_{\beta\backslash\alpha}\otimes I_\mathcal{E})T\psi h,\psi k\right\rangle\\ &= \left\langle (S^*_{\beta\backslash\alpha}\otimes I_\mathcal{E})(1\otimes h),\psi k\right\rangle=0,\end{aligned}$$

where $\beta\backslash\alpha\in\mathbb{F}_n^+$ is uniquely determined by the equation $\beta=\alpha(\beta\backslash\alpha)$. If $\beta<\alpha$, then

$$\begin{aligned}\langle T(S_\alpha\otimes I_\mathcal{E})\psi h,(S_\beta\otimes I_\mathcal{E})\psi k\rangle &= \langle T(S_{\alpha\backslash\beta}\otimes I_\mathcal{E})\psi h,\psi k\rangle\\ &= \langle (S_{\alpha\backslash\beta}\otimes I_\mathcal{E})\psi h,1\otimes k\rangle=0.\end{aligned}$$

Since T is multi-analytic, if α and β are not comparable, then

$$\langle T(S_\alpha\otimes I_\mathcal{E})\psi h,(S_\beta\otimes I_\mathcal{E})\psi k\rangle=0.$$

Therefore, if $f:=\sum_{|\alpha|\le m}e_\alpha\otimes h_\alpha$ and $g:=\sum_{|\beta|\le p}e_\beta\otimes h_\beta$ are in $F^2(H_n)\otimes\mathcal{E}$, then we have

$$\begin{aligned}\left\langle T^{1/2}M_\psi f,T^{1/2}M_\psi g\right\rangle &= \sum_{|\alpha|\le m,|\beta|\le p}\langle T(S_\alpha\otimes I_\mathcal{E})\psi h_\alpha,(S_\beta\otimes I_\mathcal{E})\psi k_\beta\rangle\\ &= \sum_{|\alpha|\le m,|\beta|\le p}\delta_{\alpha,\beta}\left\langle N^2h_\alpha,k_\beta\right\rangle\\ &= \sum_{|\alpha|\le m,|\beta|\le p}\left\langle (I\otimes N^2)(e_\alpha\otimes h_\alpha),e_\beta\otimes k_\beta\right\rangle\\ &= \left\langle (I\otimes N^2)f,g\right\rangle.\end{aligned}$$

In particular, by choosing $f=g$, we get

$$\|T^{1/2}M_\psi f\|\le\|N\|\|f\|.$$

Since $T^{1/2}$ is invertible, it follows that M_ψ can be extended to a bounded operator on $F^2(H_n)\otimes\mathcal{E}$. Therefore, M_ψ is a multi-analytic operator. The above computations, show that

$$M_\psi^*TM_\psi=I\otimes N^2. \tag{1.16}$$

Since T is a strictly positive operator, we infer that $P_\mathcal{E}T^{-1}|\mathcal{E}$ is an invertible operator on $\mathcal{E}$. The equation $P_\mathcal{E}T^{-1}|\mathcal{E}=N^2$ shows that N and $I\otimes N$ are invertible. Hence, and using relation (1.16), we deduce that M_ψ is injective. Define the map $\Lambda_0:\text{range }M_\psi\to F^2(H_n)\otimes\mathcal{E}$ by setting

$$\Lambda_0(M_\psi f):=f,\quad f\in F^2(H_n)\otimes\mathcal{E}.$$

Using relation (1.16), we have

$$\|(I\otimes N)f\|^2=\langle TM_\psi f,M_\psi f\rangle\le\|T\|\|M_\psi f\|^2.$$

On the other hand, since $I\otimes N$ is invertible, there exists a constant $K>0$ such that

$$\|(I\otimes N)f\|^2\ge K\|f\|^2=K\|\Lambda_0(M_\psi f)\|^2.$$

Combining these inequalities, we infer

$$\|\Lambda_0(M_\psi f)\|\le\sqrt{\frac{\|T\|}{K}}\|M_\psi f\|$$

for any $f \in F^2(H_n) \otimes \mathcal{E}$. Since M_ψ is outer, Λ_0 can be extended to a bounded operator on $F^2(H_n) \otimes \mathcal{E}$. It is clear that

$$\Lambda M_\psi = I. \tag{1.17}$$

On the other hand, if $g \in F^2(H_n) \otimes \mathcal{E}$ and $\{f_k\}_{k=1}^\infty$ is a sequence of elements in $F^2(H_n) \otimes \mathcal{E}$ such that $M_\psi f_k \to g$, as $k \to \infty$, then, using (1.17), we obtain

$$M_\psi \Lambda g = \lim_{k\to\infty} M_\psi \Lambda M_\psi f_k = \lim_{k\to\infty} M_\psi f_k = g.$$

Therefore, $M_\psi \Lambda = I$. Now, we can draw the conclusion that M_ψ is invertible and Λ is its inverse, which is also an outer multi-analytic operator. Moreover, relation (1.16) shows that

$$T = (M_\psi^{-1})^*(I \otimes N^2)M_\psi^{-1}.$$

Therefore, $(I \otimes N)M_\psi^{-1}$ is an outer spectral factor for T. The proof is complete. □

COROLLARY 1.6. *If* $\dim \mathcal{E} < \infty$ *and* $T \in B(F^2(H_n) \otimes \mathcal{E})$ *is a strictly positive multi-Toeplitz operator, then its entropy* $e(T)$ *satisfies the equality*

$$e(T) = -\ln \det[P_\mathcal{E} T^{-1}|\mathcal{E}].$$

PROOF. According to relations (1.13) and (1.2), we have

$$e(T) = \ln \det \Delta_T = \ln \det[P_\mathcal{E} T^{1/2} P_{\mathcal{M}_T} T^{1/2}|\mathcal{E}].$$

Therefore, it is enough to prove that

$$P_\mathcal{E} T^{1/2} P_{\mathcal{M}_T} T^{1/2}|\mathcal{E} = [P_\mathcal{E} T^{-1}|\mathcal{E}]^{-1}. \tag{1.18}$$

According to (1.7), $h \in \mathcal{M}_T$ if and only if

$$T^{1/2}h \perp (S_i \otimes I_\mathcal{E})(F^2(H_n) \otimes \mathcal{E}), \quad i = 1, \dots, n.$$

Hence, $h \in \mathcal{M}_T$ if and only if $T^{1/2}h \in \mathcal{E}$. Therefore, we have

$$\mathcal{M}_T = \text{range } (T^{-1/2}|\mathcal{E}). \tag{1.19}$$

On the other hand, it is well-known that if $X : \mathcal{X}_1 \to \mathcal{X}_2$ is an injective operator with closed range, then $X(X^*X)^{-1}X^*$ is equal to the orthogonal projection of $\mathcal{X}_2$ onto the range of X. Applying this result to the operator

$$X := (T^{-1/2}|\mathcal{E}) : \mathcal{E} \to F^2(H_n) \otimes \mathcal{E},$$

and taking into account relation (1.19), we obtain

$$P_{\mathcal{M}_T} = (T^{-1/2}|\mathcal{E})[P_\mathcal{E} T^{-1}|\mathcal{E}]^{-1}(P_\mathcal{E} T^{-1/2}).$$

This clearly implies relation (1.18). The proof is complete. □

1.2. Operators on Fock spaces and factorizations

In this section, we provide factorization results for operators in $B(\mathcal{H}_1, F^2(H_n)\otimes \mathcal{H}_2)$, where $\mathcal{H}_1$ is a finite dimensional Hilbert space, and for multi-analytic operators (see Theorem 1.7), generalizing classical results from [**29**], [**30**], [**59**], and [**67**], as well as some extensions to Fock spaces (see [**40**], [**44**], and [**48**]). A noncommutative multivariable analogue of Robinson's minimum energy delay principle (see [**57**]) for outer operators on Fock spaces is obtained (see Theorem 1.9).

A positive definite kernel on $\mathbb{F}_n^+$ is a map $K : \mathbb{F}_n^+ \times \mathbb{F}_n^+ \to B(\mathcal{H})$ with the property that

$$\sum_{i,j=1}^{k} \langle K(\sigma_i, \sigma_j)h_j, h_i\rangle \geq 0$$

for any $h_1, \ldots, h_k \in \mathcal{H}$, $\sigma_1, \ldots, \sigma_k \in \mathbb{F}_n^+$, and $k \in \mathbb{N}$. A kernel K on $\mathbb{F}_n^+$ is called multi-Toeplitz if

$$K(\sigma,\omega) := \begin{cases} K(\alpha, g_0), & \text{if } \sigma = \omega\alpha \text{ for some } \alpha \in \mathbb{F}_n^+;\\ K(g_0, \alpha), & \text{if } \omega = \sigma\alpha \text{ for some } \alpha \in \mathbb{F}_n^+;\\ 0, & \text{otherwise.}\end{cases}$$

If $K(g_0, g_0) = I_{\mathcal{H}}$, ($g_0$ is the neutral element in $\mathbb{F}_n^+$), then the kernel is called normalized. Let $\theta \in B(\mathcal{H}, F^2(H_n)\otimes \mathcal{K})$, i.e.,

$$\theta h = \sum_{\sigma\in\mathbb{F}_n^+} e_{\tilde{\sigma}} \otimes \theta_\sigma h \quad \text{for some} \quad \theta_\sigma \in B(\mathcal{H}, \mathcal{K})$$

with the property that there is $c > 0$ such that

$$\sum_{\sigma\in\mathbb{F}_n^+} \|\theta_\sigma h\|^2 \leq c\|h\|^2 \quad \text{for any} \quad h \in \mathcal{H}.$$

Denote by $\mathcal{P}(\mathcal{H})$ the set of all polynomials in $F^2(H_n)\otimes\mathcal{H}$. Define the linear operator $M_\theta : \mathcal{P}(\mathcal{H}) \to F^2(H_n)\otimes\mathcal{K}$ by $M_\theta(1\otimes h) := \theta h$ and

$$M_\theta(e_\omega \otimes h) := (S_\omega \otimes I_{\mathcal{K}})\theta h \;\text{ for any }\; h \in \mathcal{H}, \omega \in \mathbb{F}_n^+.$$

Since

$$M_\theta(S_i \otimes I_{\mathcal{H}})|\mathcal{P}(\mathcal{H}) = (S_i \otimes I_{\mathcal{K}})M_\theta|\mathcal{P}(\mathcal{H}), \quad i = 1, \ldots, n,$$

we can view M_θ as an unbounded generalized multiplier. In general, M_θ cannot be extended to a bounded linear operator from $F^2(H_n)\otimes\mathcal{H}$ to $F^2(H_n)\otimes\mathcal{K}$. However, its matrix representation

$$M_\theta := [M_{\sigma,\omega}], \quad M_{\sigma,\omega} := P_{\mathcal{K}}(S_\sigma^* \otimes I_{\mathcal{K}})M_\theta(S_\omega \otimes I_{\mathcal{H}})|_{\mathcal{H}} \in B(\mathcal{H},\mathcal{K})$$

has each column bounded as an operator from $\mathcal{H}$ to $F^2(H_n)\otimes\mathcal{K}$. It makes sense to define the kernel $K_\theta : \mathbb{F}_n^+ \times \mathbb{F}_n^+ \to B(\mathcal{H})$ by setting

$$K_\theta(\sigma,\omega) := \sum_{\alpha\in\mathbb{F}_n^+} M_{\sigma,\alpha}^* M_{\alpha,\omega},$$

where the convergence is in the SO-topology. It is easy to see that K_θ is a positive definite multi-Toeplitz kernel which is not normalized in general, i.e., $K_\theta(g_0, g_0) \neq I_{\mathcal{H}}$. Notice that if M_θ can be extended to a bounded operator, then the operator matrix $[K_\theta(\sigma,\omega)]_{\sigma,\omega\in\mathbb{F}_n^+}$ represents a multi-Toeplitz operator on $F^2(H_n)\otimes\mathcal{H}$ which is equal to $M_\theta^* M_\theta$.

In [**48**], we found an inner-outer factorization for any bounded linear operator $\theta \in B(\mathcal{H}_1, F^2(H_n) \otimes \mathcal{H}_2)$ with $K_\theta(g_0, g_0) = I$. In what follows, we show that the latter condition can be removed if $\dim \mathcal{H}_1 < \infty$ or, more generally, if the operator $K_\theta(g_0, g_0)$ has closed range. Moreover, if $\mathcal{H}_1$ is an arbitrary Hilbert space, we prove the existence of an inner-outer factorization for any bounded multi-analytic operator $M_\theta \in R_n^\infty \bar{\otimes} B(\mathcal{H}_1, \mathcal{H}_2)$. When $\mathcal{H}_1 = \mathcal{H}_2$, we obtain an explicit form of the inner-outer factorization provided in [**40**]. Our proof here is based on the existence of maximal outer factors for arbitrary multi-Toeplitz operators (see Theorem 1.1).

THEOREM 1.7. *Let $\mathcal{H}_1$ and $\mathcal{H}_2$ be Hilbert spaces.*

(i) *If $\dim \mathcal{H}_1 < \infty$, then any operator $\theta \in B(\mathcal{H}_1, F^2(H_n) \otimes \mathcal{H}_2)$ admits a factorization*
$$\theta = M_\chi \psi,$$
where $\psi \in B(\mathcal{H}_1, F^2(H_n) \otimes \mathcal{H}_3)$ is outer and $M_\chi \in R_n^\infty \bar{\otimes} B(\mathcal{H}_3, \mathcal{H}_2)$ is an inner operator. Moreover, the factorization is uniquely determined up to a diagonal unitary multi-analytic operator.

(ii) *If $\mathcal{H}_1$ is a Hilbert space and $M_\theta \in R_n^\infty \bar{\otimes} B(\mathcal{H}_1, \mathcal{H}_2)$ is a bounded multi-analytic operator, then there exist some multi-analytic operators $M_\psi \in R_n^\infty \bar{\otimes} B(\mathcal{H}_1, \mathcal{H}_3)$ and $M_\chi \in R_n^\infty \bar{\otimes} B(\mathcal{H}_3, \mathcal{H}_2)$ such that M_ψ is outer, M_χ is inner, and*
$$M_\theta = M_\chi M_\psi.$$
Moreover, the inner-outer factorization of M_θ is uniquely determined up to a unitary diagonal multi-analytic operator.

PROOF. Consider the representation
$$\theta h := \sum_{\alpha \in \mathbb{F}_n^+} e_{\tilde{\alpha}} \otimes \theta_\alpha h, \quad h \in \mathcal{H}_1,$$
where $\theta_\alpha \in B(\mathcal{H}_1, \mathcal{H}_2)$, and note that
$$X := K_\theta(g_0, g_0) = \sum_{\alpha \in \mathbb{F}_n^+} \theta_\alpha^* \theta_\alpha \in B(\mathcal{H}_1).$$
If X is an invertible operator, then the multi-Toeplitz kernel $K_{\theta X^{-1/2}}$ is normalized, i.e., $K_{\theta X^{-1/2}}(g_0, g_0) = I_{\mathcal{H}_1}$. In this case, we can apply Theorem 3.3 from [**48**] to the kernel $K_{\theta X^{-1/2}}$ and get the desired factorization.

Now, assume that X is not invertible. Let $\mathcal{N}_0 := \ker X$ and $\mathcal{H}_1 = \mathcal{N}_0 \oplus \mathcal{N}_1$ be the corresponding decomposition. For each $\alpha \in \mathbb{F}_n^+$, denote
$$\varphi_\alpha := \theta_\alpha | \mathcal{N}_1 \in B(\mathcal{N}_1, \mathcal{H}_2).$$
Note that since $\mathcal{H}_1$ is finite dimensional, the operator $\sum_{\alpha \in \mathbb{F}_n^+} \varphi_\alpha^* \varphi_\alpha \in B(\mathcal{N}_1)$ is invertible. Let us apply now the first part of the proof to the operator $\varphi \in B(\mathcal{N}_1, F^2(H_n) \otimes \mathcal{H}_2)$ defined by
$$\varphi h := \sum_{\alpha \in \mathbb{F}_n^+} e_{\tilde{\alpha}} \otimes \varphi_\alpha h, \qquad h \in \mathcal{N}_1,$$
and get the factorization
$$\varphi = M_\chi \varphi', \tag{1.20}$$

where $\varphi' \in B(\mathcal{N}_1, F^2(H_n)\otimes\mathcal{H}_3)$ is outer and $\chi \in B(\mathcal{H}_3, F^2(H_n)\otimes\mathcal{H}_2)$ is inner. If φ' has the representation

$$\varphi' h = \sum_{\alpha\in\mathbb{F}_n^+} e_{\tilde{\alpha}} \otimes \varphi'_\alpha h, \quad h \in \mathcal{N}_1,$$

with $\varphi'_\alpha \in B(\mathcal{N}_1, \mathcal{H}_3)$, define the operator $\psi \in B(\mathcal{H}, F^2(H_n)\otimes\mathcal{H}_3)$ by

$$\psi h := \sum_{\alpha\in\mathbb{F}_n^+} e_{\tilde{\alpha}} \otimes \varphi'_\alpha P_{\mathcal{N}_1} k, \quad k \in \mathcal{H}_1. \tag{1.21}$$

Since φ' is outer, it follows that ψ is also outer. On the other hand, relations (1.20), (1.21), and $\theta_\alpha|\mathcal{N}_0 = 0$, $\alpha\in\mathbb{F}_N^+$, imply $\theta = M_\chi\psi$. For the uniqueness, see the proof of Theorem 3.3 from [**48**].

Now, let us prove the second part of the theorem. Let $M_\psi \in R_n^\infty\bar{\otimes}B(\mathcal{H}_1,\mathcal{H}_3)$ be the maximal outer factor of the multi-Toeplitz operator $M_\theta^* M_\theta$. According to Theorem 1.1, we have

$$M_\theta^* M_\theta = M_\psi^* M_\psi. \tag{1.22}$$

Define the operator $Y : F^2(H_n)\otimes\mathcal{H}_3 \to F^2(H_n)\otimes\mathcal{H}_2$ by

$$Y\left[\sum_{|\sigma|\le m}(S_\sigma\otimes I_{\mathcal{H}_3})\psi h_\sigma\right] := \sum_{|\sigma|\le m}(S_\sigma\otimes I_{\mathcal{H}_2})\theta h_\sigma, \quad h_\sigma\in\mathcal{H}_1. \tag{1.23}$$

Since M_θ and M_ψ are multi-analytic operators satisfying relation (1.22) and M_ψ is outer, it is clear that Y can be extended to a unique isometry from $F^2(H_n)\otimes\mathcal{H}_3$ to $F^2(H_n)\otimes\mathcal{H}_2$. We also have $YM_\psi = M_\theta$. Due to relation (1.23), one can check that

$$Y(S_i\otimes I_{\mathcal{H}_3}) = (S_i\otimes I_{\mathcal{H}_2})Y, \quad i=1,\dots,n. \tag{1.24}$$

Indeed, for any $f\in F^2(H_n)\otimes\mathcal{H}_1$, we have

$$\begin{aligned} Y(S_i\otimes I_{\mathcal{H}_3})M_\psi f &= YM_\psi(S_i\otimes I_{\mathcal{H}_3})f = M_\theta(S_i\otimes I_{\mathcal{H}_1})f \\ &= (S_i\otimes I_{\mathcal{H}_1})M_\theta f = (S_i\otimes I_{\mathcal{H}_2})YM_\psi f. \end{aligned}$$

Since M_ψ is outer, relation (1.24) follows. According to [**44**], there exists an inner multi-analytic operator $M_\chi\in R_n^\infty\bar{\otimes}B(\mathcal{H}_3,\mathcal{H}_2)$ such that $Y = M_\psi$. Therefore, we have $M_\theta = M_\chi M_\psi$, which is the desired inner-outer factorization.

To prove the uniqueness, let $M_\theta = M_{\chi'}M_{\psi'}$ be another inner-outer factorization with $M_{\chi'}\in R_n^\infty\bar{\otimes}B(\mathcal{H}_3',\mathcal{H}_2)$ inner and $M_{\psi'}\in R_n^\infty\bar{\otimes}B(\mathcal{H}_1,\mathcal{H}_3')$ outer. Then we have

$$M_\theta^* M_\theta = M_\psi^* M_\psi = M_{\psi'}^* M_{\psi'}.$$

Now, we can find an inner multi-analytic operator $Z\in R_n^\infty\bar{\otimes}B(\mathcal{H}_3,\mathcal{H}_3')$ such that $ZM_\psi = M_{\psi'}$. Since $M_{\psi'}$ is outer, we deduce that Z is unitary. According to [**39**], $Z = I\otimes U$ where $U\in B(\mathcal{H}_3,\mathcal{H}_3')$ is unitary. Now the equation $M_\chi M_\psi = M_{\chi'}M_{\psi'}$ implies $M_\chi M_\psi = M_{\chi'}ZM_\psi$. Since M_ψ is outer, we obtain $M_\chi = M_{\chi'}Z$. This completes the proof. □

REMARK 1.8. *Theorem 1.7 part* (i) *remains true if* $\mathcal{H}_1$ *is an arbitrary Hilbert space provided that the operator* $K_\theta(g_0,g_0)$ *has closed range.*

Given a Hilbert space $\mathcal{E}$, let $\mathcal{P}_k(\mathcal{E}) := \bigoplus_{|\alpha|\leq k} e_\alpha \otimes \mathcal{E}$ be the set of all polynomials of degree $\leq k$. For any Hilbert space $\mathcal{E}$, denote by $\mathbb{P}_k$ the orthogonal projection of $F^2(H_n)\otimes\mathcal{E}$ onto $\mathcal{P}_k(\mathcal{E})$. We say that $\theta \in B(\mathcal{E}, F^2(H_n)\otimes\mathcal{G})$ is outer up to a constant inner operator on the left if θ admits a factorization $\theta = M_\chi\psi$ where $\psi \in B(\mathcal{E}, F^2(H_n)\otimes\mathcal{E}_1)$ is outer and $M_\chi \in R_n^\infty\bar{\otimes}B(\mathcal{E}_1,\mathcal{G})$ is a constant inner operator, i.e., $M_\chi = I\otimes V$, where $V\in B(\mathcal{E},\mathcal{G})$ is an isometry. Note that if $M_\theta \in R_n^\infty\bar{\otimes}B(\mathcal{E},\mathcal{G})$ then $M_\psi \in R_n^\infty\bar{\otimes}B(\mathcal{E},\mathcal{E}_1)$.

The next theorem is a noncommutative multivariable analogue of Robinson's minimum energy delay principle for outer functions (see [**57**], [**22**]).

THEOREM 1.9. *Let k be a fixed nonnegative integer.*

(i) *If $\mathcal{E}$ be a finite dimensional space, then an operator $\theta\in B(\mathcal{E}, F^2(H_n)\otimes\mathcal{G})$ is outer up to a constant inner operator on the left if and only if*

$$\|\mathbb{P}_k M_\psi p\| \leq \|\mathbb{P}_k M_\theta p\|, \quad p\in\mathcal{P}_k(\mathcal{E}), \tag{1.25}$$

for any operator $\psi\in B(\mathcal{E}, F^2(H_n)\otimes\mathcal{Y})$ satisfying $K_\psi = K_\theta$.

(ii) *If $\mathcal{E}$ is an arbitrary Hilbert space, then an operator $M_\theta\in R_n^\infty\bar{\otimes}B(\mathcal{E},\mathcal{G})$ is outer up to a constant inner operator on the left if and only if the inequality* (1.25) *holds for any multi-analytic operator $M_\psi\in R_n^\infty\bar{\otimes}B(\mathcal{E},\mathcal{Y})$ satisfying $M_\psi^* M_\psi = M_\theta^* M_\theta$.*

PROOF. Assume that $\theta\in B(\mathcal{E}, F^2(H_n)\otimes\mathcal{G})$ is outer and $\psi\in B(\mathcal{E}, F^2(H_n)\otimes\mathcal{Y})$ is an operator satisfying $K_\psi = K_\theta$. Define $Q: F^2(H_n)\otimes\mathcal{G}\to F^2(H_n)\otimes\mathcal{Y}$ by

$$Q\Big(\sum_{|\sigma|\leq m}(S_\sigma\otimes I_\mathcal{G})\theta h_\sigma\Big) := \sum_{|\sigma|\leq m}(S_\sigma\otimes I_\mathcal{Y})\psi h_\sigma, \qquad h_\sigma\in\mathcal{E}.$$

Since $K_\psi = K_\theta$ and θ is an outer operator, Q extends to an isometry from $F^2(H_n)\otimes\mathcal{G}$ to $F^2(H_n)\otimes\mathcal{Y}$ such that $QM_\theta|\mathcal{P}(\mathcal{E}) = M_\psi|\mathcal{P}(\mathcal{E})$. Since

$$Q(S_i\otimes I_\mathcal{G}) = (S_i\otimes I_\mathcal{Y})Q, \qquad i=1,\dots,n,$$

Q is an inner multi-analytic operator. Therefore, $M_\psi|\mathcal{P}(\mathcal{E}) = M_f M_\theta|\mathcal{P}(\mathcal{E})$, where $M_f\in R_n^\infty\bar{\otimes}B(\mathcal{G},\mathcal{Y})$ is an inner operator. Note that

$$\mathbb{P}_k M_\theta|\mathcal{P}(\mathcal{E}) = \mathbb{P}_k M_\theta \mathbb{P}_k|\mathcal{P}(\mathcal{E}) \quad \text{and} \quad \mathbb{P}_k M_f = \mathbb{P}_k M_f\mathbb{P}_k.$$

Consequently, for any $p\in\mathcal{P}_k(\mathcal{E})$, we have

$$\|\mathbb{P}_k M_\psi p\| = \|\mathbb{P}_k M_f M_\theta p\| = \|\mathbb{P}_k M_f \mathbb{P}_k M_\theta p\| \leq \|\mathbb{P}_k M_\theta p\|. \tag{1.26}$$

Hence, we deduce relation (1.25).

Conversely, assume that $\theta\in B(\mathcal{E}, F^2(H_n)\otimes\mathcal{G})$ is an operator such that relation (1.25) holds. According to Theorem 1.7 part (i), we have $\theta = M_\chi\varphi$, where $\varphi\in B(\mathcal{E}, F^2(H_n)\otimes\mathcal{E}_1)$ is outer and $M_\chi\in R_n^\infty\bar{\otimes}B(\mathcal{E}_1,\mathcal{G})$ is inner. Since $K_\theta = K_\varphi$, we can apply the first part of the proof to the outer operator φ, to infer that

$$\|\mathbb{P}_k M_\theta p\| \leq \|\mathbb{P}_k M_\varphi p\|, \quad p\in\mathcal{P}_k(\mathcal{E}). \tag{1.27}$$

Combining relation (1.25) (when $\psi=\varphi$) with relation (1.27), we obtain

$$\|\mathbb{P}_k M_\theta p\| = \|\mathbb{P}_k M_\varphi p\|, \quad p\in\mathcal{P}_k(\mathcal{E}). \tag{1.28}$$

Since $\mathbb{P}_k M_\chi = \mathbb{P}_k M_\chi \mathbb{P}_k$ and M_χ is inner, we have

$$\|\mathbb{P}_k M_\theta p\| = \|\mathbb{P}_k M_\chi M_\varphi p\| = \|\mathbb{P}_k M_\chi \mathbb{P}_k M_\varphi p\| \leq \|\mathbb{P}_k M_\varphi p\|, \quad p\in\mathcal{P}(\mathcal{E}).$$

On the other hand, since φ is outer, we can use relation (1.28) to deduce that $\mathbb{P}_k M_\chi \mathbb{P}_k$ is an isometry from $\mathcal{P}_k(\mathcal{E}_1)$ to $\mathcal{P}_k(\mathcal{E})$. Let

$$M_\chi = \sum_{\alpha\in\mathbb{F}_n^+} R_\alpha \otimes \chi_{(\alpha)}, \quad \chi_{(\alpha)} \in B(\mathcal{E}_1, \mathcal{G}),$$

be the Fourier representation of the multi-analytic operator M_χ, and let $x \in \mathcal{E}_1$. Note that if $\alpha \in \mathbb{F}_n^+$, $|\alpha| = k$, then

$$\mathbb{P}_k M_\chi (S_\alpha \otimes I_{\mathcal{E}_1})(1 \otimes x) = \mathbb{P}_k (S_\alpha \otimes I_{\mathcal{G}}) M_\chi (1 \otimes x) = e_\alpha \otimes \chi_{(0)} x.$$

Since $\mathbb{P}_k M_\chi \mathbb{P}_k$ is an isometry from $\mathcal{P}_k(\mathcal{E}_1)$ to $\mathcal{P}_k(\mathcal{E})$, we have

$$\|\chi_{(0)} x\| = \|\mathbb{P}_k M_\chi (S_\alpha \otimes I_{\mathcal{E}_1})(1 \otimes x)\| = \|(S_\alpha \otimes I_{\mathcal{E}_1})(1 \otimes x)\| = \|x\|$$

for any $x \in \mathcal{E}_1$. Thus $\chi_{(0)}$ is an isometry from $\mathcal{E}_1$ to $\mathcal{G}$. Moreover, since M_χ is inner, we deduce that

$$\|\chi_{(0)} x\|^2 = \|x\|^2 = \|M_\chi(1 \otimes x)\|^2 = \sum_{\alpha\in\mathbb{F}_n^+} \|\chi_{(\alpha)} x\|^2$$

for any $x \in \mathcal{E}_1$. Hence, $M_\chi = I \otimes \chi_{(0)}$, i.e., a constant inner operator.

To prove part (ii) of the theorem, note that if $M_\theta \in R_n^\infty \bar{\otimes} B(\mathcal{E}, \mathcal{G})$, then $[K_\theta(\sigma, \omega)]_{\sigma,\omega\in\mathbb{F}_n^+}$ is the matrix representation of the multi-Toeplitz operator $M_\theta^* M_\theta$. The proof of part (ii) is very similar to the proof of part (i). The only difference is that, in this case, we have to use Theorem 1.7 part (ii) for the inner-outer factorization of M_θ. □

Let us mention that, using Remark 1.8, one can show that Theorem 1.9 part (i) remains true if $\mathcal{H}_1$ is an arbitrary Hilbert space and the operator $K_\theta(g_0, g_0)$ has closed range.

1.3. Prediction entropy for positive definite multi-Toeplitz kernels on free semigroups

In this section, we define the notion of prediction entropy and prove entropy inequalities for positive definite multi-Toeplitz kernels on the free semigroup $\mathbb{F}_n^+$ and multi-analytic operators (see Theorem 1.10). We extend the classical result (see [**31**]) which stated for $H^2(\mathbb{D})$ says that if $f \in H^2(\mathbb{D})$, then $\ln|f(e^{it})|$ is integrable and

$$\frac{1}{2\pi}\int_{-\pi}^{\pi} \ln|f(e^{it})|\, dt \geq \ln|f(0)|.$$

Next, we give a characterization for the outer operators in $B(\mathcal{E}, F^2(H_n) \otimes \mathcal{E})$ if $\dim \mathcal{E} < \infty$. In particular, we find a noncommutative multivariable analogue of the following classical result. A function $f \in H^2(\mathbb{D})$ is outer if and only if $f(0) \neq 0$ and

$$\frac{1}{2\pi}\int_{-\pi}^{\pi} \ln|f(e^{it})|\, dt = \ln|f(0)|.$$

Let $K : \mathbb{F}_n^+ \times \mathbb{F}_n^+ \to B(\mathcal{E})$ be a positive definite multi-Toeplitz kernel (not necessarily normalized). Define the positive operator $\Delta_K \in B(\mathcal{E})$ by setting

$$\langle \Delta_K h, h\rangle := \inf_{h_{g_0}=h,\ h_\sigma\in\mathcal{E}} \sum \langle K(\omega,\sigma) h_\sigma, h_\omega\rangle, \quad h \in \mathcal{E}, \tag{1.29}$$

where the sum is taken over all finitely supported sequences $\{h_\sigma\}_{\sigma\in\mathbb{F}_n^+} \subset \mathcal{E}$. When $\dim \mathcal{E} < \infty$, we define the prediction entropy of the kernel K by

$$e(K) := \ln \det \Delta_K. \tag{1.30}$$

Let $\mathcal{T} := [T_1 \ \cdots \ T_n]$, $T_i \in B(\mathcal{H})$, and define the kernel $K_\mathcal{T} : \mathbb{F}_n^+ \times \mathbb{F}_n^+ \to B(\mathcal{H})$ by setting

$$K_\mathcal{T}(\sigma,\omega) := \begin{cases} T_\alpha, & \text{if } \omega = \sigma\alpha \text{ for some } \alpha \in \mathbb{F}_n^+; \\ T_\alpha^*, & \text{if } \sigma = \omega\alpha \text{ for some } \alpha \in \mathbb{F}_n^+; \\ 0, & \text{otherwise.} \end{cases}$$

We proved in [**48**] that the multi-Toeplitz kernel $K_\mathcal{T}$ is positive definite if and only if $[T_1 \ \cdots \ T_n]$ is a row contraction. In this case, if $\dim \mathcal{H} < \infty$, we can use Theorem 1.3 and Theorem 4.1 from [**48**] to deduce that the entropy of $K_\mathcal{T}$ satisfies

$$e(K_\mathcal{T}) = \ln \det(I - T_1 T_n^* - \cdots - T_n T_n^*).$$

In what follows, we provide entropy inequalities for positive definite multi-Toeplitz kernels on free semigroups and multi-Toeplitz operators on Fock spaces.

THEOREM 1.10. *Let $\mathcal{E}$ be a Hilbert space and $\theta \in B(\mathcal{E}, F^2(H_n)\otimes\mathcal{E})$ be a nonzero operator.*

(i) *If $\dim \mathcal{E} < \infty$, then $\Delta_{K_\theta} \neq 0$ and*

$$\langle \Delta_{K_\theta} h, h\rangle \geq \sum_{|\alpha|=k} \langle \theta_\alpha^* \theta_\alpha h, h\rangle\,, \tag{1.31}$$

where k is the smallest nonnegative integer such that $\theta_{\alpha_0} \neq 0$ for some $\alpha_0 \in \mathbb{F}_n^+$ with $|\alpha_0| = k$. Moreover, the entropy of the multi-Toeplitz kernel K_θ satisfies the inequality

$$e(K_\theta) \geq \ln \det \left[\sum_{|\alpha|=k} \theta_\alpha^* \theta_\alpha \right]. \tag{1.32}$$

(ii) *If $\mathcal{E}$ is an arbitrary Hilbert space and $M_\theta \in R_n^\infty \bar{\otimes} B(\mathcal{E})$ is a nonzero multi-analytic operator, then inequality* (1.31) *remains true. If, in addition, $\dim \mathcal{E} < \infty$, then inequality* (1.32) *holds.*

In particular, if $\theta \in F^2(H_n)$, then $e(K_\theta) > -\infty$ and $e(K_\theta) \geq \ln |\theta(0)|^2$.

PROOF. Consider the Fourier representation

$$\theta h := \sum_{\alpha\in\mathbb{F}_n^+} e_{\tilde\alpha} \otimes \theta_\alpha h, \quad h \in \mathcal{E},$$

and assume that $\theta(0) \neq 0$ (recall that $\theta(0) := \theta_{g_0} \in B(\mathcal{E})$). Since $\dim \mathcal{E} < \infty$, there is an invertible operator $X : \ker \theta(0) \to \ker \theta(0)^*$. The operator $\theta^\epsilon(0) \in B(\mathcal{E})$ defined by

$$\theta^\epsilon(0) := \theta(0) + \epsilon X P_{\ker \theta(0)}$$

is invertible for any $\epsilon > 0$. Define $\theta^\epsilon : \mathcal{E} \to F^2(H_n) \otimes \mathcal{E}$ by

$$\theta^\epsilon h := 1 \otimes \theta^\epsilon(0) h + \sum_{|\alpha|\geq 1} e_{\tilde\alpha} \otimes \theta_\alpha h, \quad h \in \mathcal{E}. \tag{1.33}$$

Since $N := \sum_{\alpha\in\mathbb{F}_n^+} \theta_\alpha^\epsilon{}^*\theta_\alpha^\epsilon \in B(\mathcal{E})$ is invertible, we can define $\psi : \mathcal{E} \to F^2(H_n)\otimes\mathcal{E}$ by setting

$$\psi h := \sum_{\alpha\in\mathbb{F}_n^+} e_{\tilde\alpha}\otimes \theta_\alpha^\epsilon N^{-1/2}h, \quad h\in\mathcal{E}.$$

Applying Theorem 1.7 part (i), we find an outer operator $\varphi : \mathcal{E} \to F^2(H_n)\otimes\mathcal{E}_3$ such that $K_\psi = K_\varphi$. Since

$$K_\psi(g_0, g_0) = \sum_{\alpha\in\mathbb{F}_n^+} \psi_\alpha^*\psi_\alpha = I_{\mathcal{E}},$$

we can apply Theorem 4.1 from [**48**] to the normalized multi-Toeplitz kernel K_ψ and deduce the equality

$$\inf_{h_{g_0}=h,\ h_\sigma\in\mathcal{E}} \sum \langle K_\psi(\omega,\sigma)h_\sigma, h_\omega\rangle = \langle \varphi(0)^*\varphi(0)h, h\rangle, \quad h\in\mathcal{E}, \tag{1.34}$$

where the sum is taken over all finitely supported sequences $\{h_\sigma\}_{\sigma\in\mathbb{F}_n^+}\subset\mathcal{E}$. Since $K_\psi = K_\varphi$, we can apply Theorem 1.9 part (i) when $k=0$, to infer that

$$\langle \varphi(0)^*\varphi(0)h, h\rangle \geq \langle \psi(0)^*\psi(0)h, h\rangle, \quad h\in\mathcal{E}.$$

This inequality together with equation (1.34) imply

$$\Delta_{K_\psi} \geq \psi(0)^*\psi(0). \tag{1.35}$$

Since $K_\psi(\sigma,\omega) = N^{-1/2}K_\theta(\sigma,\omega)N^{-1/2}$, relation (1.35) implies

$$\Delta_{K_{\theta^\epsilon}} \geq \theta^\epsilon(0)^*\theta^\epsilon(0) \tag{1.36}$$

for any $\epsilon > 0$. Note that

$$\langle \Delta_{K_{\theta^\epsilon}}h, h\rangle = \inf_{p\in\mathcal{P}(\mathcal{E}),\ p(0)=0} \|(I\otimes \epsilon X P_{\ker\theta(0)} + M_\theta)(h-p)\|^2.$$

Taking $\epsilon\to 0$ in inequality (1.36), we get

$$\Delta_{K_\theta} \geq \theta(0)^*\theta(0).$$

Now, assume that $\theta(0) = 0$ and let k is the smallest nonnegative integer such that $\theta_{\alpha_0}\neq 0$ for some $\alpha_0\in\mathbb{F}_n^+$ with $|\alpha_0| = k$. Then we have

$$\theta = \sum_{|\alpha|\geq k} e_{\tilde\alpha}\otimes\theta_\alpha = \sum_{|\beta|=k}(R_\beta\otimes I)\Lambda_\beta$$

for some operators $\Lambda_\beta : \mathcal{E}\to F^2(H_n)\otimes\mathcal{E}$. Note that $\Lambda_\beta(0) = \theta_\beta$, $\beta\in\mathbb{F}_n^+$. Since $\{R_\beta\}_{|\beta|=k}$ are isometries with orthogonal ranges, we deduce

$$\|M_\theta(h-p)\|^2 = \sum_{|\beta|=k}\|M_{\Lambda_\beta}(h-p)\|^2$$

for any $h\in\mathcal{E}$ and $p\in\mathcal{P}(\mathcal{E})$ with $p(0)=0$. Therefore, applying the first part of the proof to each operator Λ_β, $\beta\in\mathbb{F}_n^+$ with $|\beta| = k$, we have

$$\begin{aligned}\inf_{p\in\mathcal{P}(\mathcal{E}),\ p(0)=0}\|M_\theta(h-p)\|^2 &\geq \sum_{|\beta|=k}\inf_{p\in\mathcal{P}(\mathcal{E}),\ p(0)=0}\|M_{\Lambda_\beta}(h-p)\|^2\\ &\geq \sum_{|\beta|=k}\langle\Lambda_\beta(0)^*\Lambda_\beta(0)h, h\rangle\end{aligned}$$

for any $h \in \mathcal{E}$. Since $\Lambda_{\alpha_0}(0) = \theta_{\alpha_0} \neq 0$ and

$$\langle \Delta_{K_\theta} h, h \rangle = \inf_{p \in \mathcal{P}(\mathcal{E}),\ p(0)=0} \|M_\theta(h-p)\|^2,$$

we deduce that $\Delta_{K_\theta} \neq 0$ and relation (1.31) is proved. The inequality (1.32) is now obvious.

To prove part (ii) of the theorem, let $\mathcal{E}$ be an arbitrary Hilbert space and let $M_\theta \in R_n^\infty \bar{\otimes} B(\mathcal{E})$ be a nonzero multi-analytic operator such that $\theta(0) \neq 0$. Using the Szegö type result of Theorem 1.3 when $T = M_\theta^* M_\theta$, we obtain

$$\inf_{p \in \mathcal{P}(\mathcal{E}),\ p(0)=0} \langle M_\theta^* M_\theta (h-p), h-p \rangle = \langle \varphi(0)^* \varphi(0) h, h \rangle, \tag{1.37}$$

where M_φ is the maximal outer factor of T. Moreover, due to Theorem 1.1, we have $M_\theta^* M_\theta = M_\varphi^* M_\varphi$. Now, Theorem 1.9 part (ii) implies

$$\langle \varphi(0)^* \varphi(0) h, h \rangle \geq \langle \theta(0)^* \theta(0) h, h \rangle \tag{1.38}$$

for any $h \in \mathcal{E}$. Combining relations (1.37) and (1.38), we get

$$\Delta_{M_\theta^* M_\theta} \geq \theta(0)^* \theta(0).$$

When $\theta(0) = 0$, the proof is the same as that of part (i) of the theorem. Noting that $\Delta_{K_\theta} = \Delta_{M_\theta^* M_\theta}$ and $e(K_\theta) = e(M_\theta^* M_\theta)$, one can easily complete the proof. □

Let us remark that if $\mathcal{E}$ is an arbitrary Hilbert space and $\theta \in B(\mathcal{E}, F^2(H_n) \otimes \mathcal{E})$ is such that the operator $K_\theta(g_0, g_0)$ has closed range, then the results of Theorem 1.10 part (i) remain true.

REMARK 1.11. *If $f \in F^2(H_1)$, then*

$$e(K_f) = \frac{1}{2\pi} \int_{-\pi}^{\pi} \ln |f(e^{it})|^2 \, dt. \tag{1.39}$$

PROOF. Under the canonical identification of the full Fock space $F^2(H_1)$ with the Hardy space $H^2(\mathbb{D})$, we have

$$\begin{aligned} \inf \|M_f(1-p)\|^2 &= \inf \|f(1-p)\|^2 \\ &= \inf \left\{ \frac{1}{2\pi} \int_{-\pi}^{\pi} |1 - p(e^{it})|^2 |f(e^{it})|^2 \, dt \right\} \\ &= \exp \left[\frac{1}{2\pi} \int_{-\pi}^{\pi} \ln |f(e^{it})|^2 \, dt \right], \end{aligned}$$

where the infimum is taken over all polynomials $p \in F^2(H_1)$ with $p(0) = 0$. The latter equality is due to Szegö's theorem [**31**]. According to relation (1.29) and (1.30), thc equality (1.39) follows. □

A characterization for the outer operators in $B(\mathcal{E}, F^2(H_n) \otimes \mathcal{E})$, when $\mathcal{E}$ is finite dimensional, is proved in what follows.

THEOREM 1.12. *Let k be a fixed nonnegative integer, $\mathcal{E}$ be a finite dimensional Hilbert space, and let $\theta \in B(\mathcal{E}, F^2(H_n) \otimes \mathcal{E})$. Then the following statements are equivalent:*

(i) *θ is an outer operator;*

(ii) $\theta(0)$ *is invertible and if* $\psi \in B(\mathcal{E}, F^2(H_n)\otimes\mathcal{E})$ *is such that* $K_\psi = K_\theta$, *then*

$$\|\mathbb{P}_k M_\psi p\| \leq \|\mathbb{P}_k M_\theta p\|, \quad p \in \mathcal{P}_k(\mathcal{E});$$

(iii) $\theta(0)$ *is invertible and*

$$\inf_{h_\sigma\in\mathcal{E},\ h_e=h} \sum \langle K_\theta(\omega,\sigma)h_\sigma, h_\omega\rangle = \langle\theta(0)^*\theta(0)h, h\rangle\,, \quad h \in \mathcal{E}, \tag{1.40}$$

where the sum is taken over all finitely supported sequences $\{h_\sigma\}_{\sigma\in\mathbb{F}_n^+} \subset \mathcal{E}$ *such that* $h_e = h$;

(iv) $\theta(0)$ *is invertible and the entropy of the positive definite multi-Toeplitz kernel* K_θ *satisfies the equation*

$$e(K_\theta) = \ln\det[\theta(0)^*\theta(0)].$$

PROOF. Assume that θ is an outer operator. First, we show that the operator $\theta(0) := P_\mathcal{E}\theta \in B(\mathcal{E})$ is invertible. Suppose that there exists $y \in \mathcal{E}$, $y \neq 0$, such that y is orthogonal to the range of $\theta(0)$. Note that, for any $h \in \mathcal{E}$, $\alpha \in \mathbb{F}_n^+$, we have

$$\langle (S_\alpha \otimes I_\mathcal{E})\theta h, y\rangle = 0, \quad \text{if } |\alpha| \geq 1,$$

and

$$\langle\theta h, y\rangle = \langle\theta(0)h, y\rangle = 0.$$

Hence, y is orthogonal to the linear span $\bigvee_{\sigma\in\mathbb{F}_n^+}(S_\sigma \otimes I_\mathcal{E})\theta\mathcal{E}$, which contradicts that θ is outer. Therefore, the range of $\theta(0)$ is dense in $\mathcal{E}$. Since $\dim\mathcal{E} < \infty$, it is clear that $\theta(0)$ must be invertible. Now, the implication (i) $\Longrightarrow$ (ii) follows from Theorem 1.9 part (i). Assume (ii) holds. According to the same theorem, the operator θ has a factorization

$$\theta = M_\chi\psi_0 \tag{1.41}$$

with $\psi_0 \in B(\mathcal{E}, F^2(H_n)\otimes\mathcal{E}_1)$ outer and $M_\chi = I_{F^2(H_n)} \otimes V$, where $V \in B(\mathcal{E}_1, \mathcal{E})$ is an isometry. This shows that $\dim\mathcal{E}_1 \leq \dim\mathcal{E}$. Since ψ_0 is outer, the range of $\psi_0(0)$ is dense in $\mathcal{E}_1$. On the other hand, since $\mathcal{E}$ is finite dimensional, V is an isometry, and $\theta(0) = V\psi_0(0)$ is invertible, it follows that $\dim\mathcal{E}_1 = \dim\mathcal{E}$. Using relation (1.41), we deduce that θ is outer. Therefore (ii)$\Leftrightarrow$ (i).

Assume now that (i) holds. Since the operator $X := \sum_{\alpha\in\mathbb{F}_n^+} \theta_\alpha^*\theta_\alpha \in B(\mathcal{E})$ is invertible, $\theta X^{-1/2} \in B(\mathcal{E}, F^2(H_n)\otimes\mathcal{E})$ is outer and $K_{\theta X^{-1/2}}(g_0, g_0) = I$. Consequently, applying Theorem 4.1 from [**48**], we get

$$\inf_{h_{g_0}=h,\ h_\sigma\in\mathcal{E}} \sum \langle K_{\theta X^{-1/2}}(\omega,\sigma)h_\sigma, h_\omega\rangle = \left\langle\theta(0)^*\theta(0)X^{-1/2}h, X^{-1/2}h\right\rangle, \quad h \in \mathcal{E},$$

where the sum is taken over all finitely supported sequences $\{h_\sigma\}_{\sigma\in\mathbb{F}_n^+} \subset \mathcal{E}$. Hence, (iii) follows. Assume now that (iii) holds. Let $\varphi : \mathcal{E} \to F^2(H_n)\otimes\mathcal{E}_1$ be the maximal outer factor of the multi-Toeplitz kernel $K_{\theta X^{-1/2}}$. According to Theorem 3.3 from [**48**], we have $K_{\theta X^{-1/2}} = K_\varphi$. Using relation (1.40) and Theorem 4.1 from [**48**], we obtain

$$\begin{aligned}\left\langle\theta(0)^*\theta(0)X^{-1/2}h, X^{-1/2}h\right\rangle &= \inf_{h_{g_0}=h,\ h_\sigma\in\mathcal{E}} \sum \langle K_{\theta X^{-1/2}}(\omega,\sigma)h_\sigma, h_\omega\rangle \\ &= \langle\varphi(0)^*\varphi(0)h, h\rangle\end{aligned}$$

for any $h \in \mathcal{E}$. Now, as in the proof of Theorem 1.9 (when $k = 0$), we obtain the factorization

$$\theta X^{-1/2} = M_\chi \varphi,$$

where $M_\chi = I_{F^2(H_n)} \otimes W$ and $W \in B(\mathcal{E}_1, \mathcal{E})$ is an isometry. Since θ and X are invertible and $\theta(0)X^{-1/2} = W\varphi(0)$, we must have $W\mathcal{E}_1 = \mathcal{E}$. Thus, M_χ is a unitary operator. Since φ is outer, it is clear that $\theta = M_\chi \varphi X^{1/2}$ is outer, i.e., (i) holds. Therefore (iii)$\Leftrightarrow$ (i).

To prove the equivalence (iii) $\Leftrightarrow$ (iv) is enough to show that if $\theta(0)$ is invertible, then $\Delta_{K_\theta} = \theta(0)^*\theta(0)$ if and only if

$$\det \Delta_{K_\theta} = \det[\theta(0)^*\theta(0)]. \tag{1.42}$$

Indeed, according to Theorem 1.10, we have

$$\Delta_{K_\theta} \geq \theta(0)^*\theta(0).$$

Hence, it follows that there is a contraction $C \in B(\mathcal{E})$ such that

$$\theta(0)^*\theta(0) = \Delta_{K_\theta}^{1/2} C^* C \Delta_{K_\theta}^{1/2}, \tag{1.43}$$

whence we deduce that

$$\det[\theta(0)^*\theta(0)] = \det \Delta_{K_\theta} \det(C^*C). \tag{1.44}$$

Since $\theta(0)$ is an invertible operator, one can see that relations (1.42) and (1.44) imply that $\det(C^*C) = 1$. Hence, and using the fact that C is a contraction, we get $C^*C = I$. Thus, relation (1.43) becomes $\theta(0)^*\theta(0) = \Delta_{K_\theta}$, which proves our claim. The proof is complete. □

COROLLARY 1.13. *If $f \in F^2(H_n)$, then f is outer if and only if $f(0) \neq 0$ and*

$$e(K_f) = \ln |f(0)|^2.$$

Note that in the particular case when $n = 1$ and $\mathcal{E} = \mathbb{C}$, Theorem 1.12 and Remark 1.11 imply the following classical result. A function $f \in H^2(\mathbb{D})$ is outer if and only if $f(0) \neq 0$ and

$$\frac{1}{2\pi} \int_{-\pi}^{\pi} \ln |f(e^{it})| \, dt = \ln |f(0)|.$$

1.4. Extreme points of the unit ball of F_n^∞

It is well-known [**31**] that a function $f \in H^\infty(\mathbb{D})$ is an extreme point of the unit ball of $H^\infty(\mathbb{D})$ if and only if

$$\frac{1}{2\pi} \int_{-\pi}^{\pi} \ln(1 - |f(e^{it})|^2) \, dt = -\infty.$$

In what follows, we present some results concerning the extreme points of the unit ball of the noncommutative analytic Toeplitz algebra F_n^∞ and $F_n^\infty \bar{\otimes} B(\mathcal{H})$, where $\dim \mathcal{H} < \infty$. In particular, we prove that if $\varphi \in F_n^\infty$, $\|\varphi\| \leq 1$, and the entropy $E(\varphi) = -\infty$, then φ is an extreme point of the unit ball of F_n^∞. For the converse, a weaker form is provided.

Let $[V_1' \cdots V_n']$, $V_i' \in B(\mathcal{K}')$, be isometries with orthogonal ranges on a Hilbert space $\mathcal{K}'$ and let $B : F^2(H_n) \otimes \mathcal{E} \to \mathcal{K}'$ be a contractive generalized multiplier with respect to $\{S_i \otimes I_{\mathcal{E}}\}_{i=1}^n$ and $\{V_i'\}_{i=1}^n$, i.e.,

$$B(S_i \otimes I_{\mathcal{E}}) = V_i' B, \quad i = 1, \ldots n.$$

Note that

$$\begin{aligned}(S_i^* \otimes I_{\mathcal{E}})(I_{F^2(H_n)\otimes\mathcal{E}} - B^*B)(S_j \otimes I_{\mathcal{E}}) &= \delta_{ij} I_{F^2(H_n)\otimes\mathcal{E}} - B^* V_i'^* V_j B \\ &= \delta_{ij}(I_{F^2(H_n)\otimes\mathcal{E}} - B^*B)\end{aligned}$$

for any $i, j = 1, \ldots, n$. Therefore, D_B^2 is a multi-Toeplitz operator on $F^2(H_n) \otimes \mathcal{E}$. If $\dim \mathcal{E} < \infty$, we define the prediction entropy of the generalized multiplier B by setting

$$E(B) := e(D_B^2). \tag{1.45}$$

In particular, if $\mathcal{K}' := F^2(H_n) \otimes \mathcal{K}$ and $V_i' := S_i \otimes I_{\mathcal{K}}$, $i = 1, \ldots, n$, then the generalized multipliers B are multi-analytic operators on Fock spaces. Let $\Theta : F^2(H_n) \otimes \mathcal{E} \to F^2(H_n) \otimes \mathcal{K}$ be a multi-analytic operator, i.e., $\Theta \in R_n^\infty \bar{\otimes} B(\mathcal{E}, \mathcal{K})$. If $\|\Theta\| \leq 1$ and $\dim \mathcal{E} < \infty$, then the prediction entropy of the multi-analytic operator Θ satisfies the equation

$$E(\Theta) = \ln \det \Delta(\Theta),$$

where

$$\langle \Delta(\Theta)x, x\rangle := \inf\{\langle (I - \Theta^*\Theta)(x - p), x - p\rangle : \ p \in F^2(H_n) \otimes \mathcal{E}, \ p(0) = 0\} \tag{1.46}$$

for any $x \in \mathcal{E}$. Using Szegö's theorem in the particular case when $n = 1$ and $\mathcal{E} = \mathcal{K} = \mathbb{C}$, we have

$$E(f) = \frac{1}{2\pi} \int_{-\pi}^{\pi} \ln(1 - |f(e^{it})|^2)\, dt,$$

which is the classical definition of the entropy of $f \in H^\infty(\mathbb{D})$ with $\|f\| \leq 1$.

Since $R_n^\infty = U^* F_n^\infty U$, where U is the flipping operator, all the results of this section are true for both algebras F_n^∞ and R_n^∞.

THEOREM 1.14. *Let $M_\theta \in R_n^\infty \bar{\otimes} B(\mathcal{E})$ be such that $\|M_\theta\| \leq 1$.*

(i) *If $\Delta(M_\theta) = 0$, then M_θ is an extreme point of the unit ball of $R_n^\infty \bar{\otimes} B(\mathcal{E})$.*
(ii) *Assume that there is a multi-analytic operator $M_\varphi \in R_n^\infty \bar{\otimes} B(\mathcal{E})$, $M_\varphi \neq 0$, such that its range is orthogonal to the range of M_θ. If $\dim \mathcal{E} < \infty$ and the entropy $E(M_\theta) > -\infty$, then M_θ is not an extreme point of the unit ball of $R_n^\infty \bar{\otimes} B(\mathcal{E})$.*

PROOF. Assume that $\Delta(M_\theta) = 0$, and let $G \in R_n^\infty \bar{\otimes} B(\mathcal{E})$ be such that

$$\|M_\theta + G\| \leq 1 \quad \text{and} \quad \|M_\theta - G\| \leq 1.$$

Hence, we get

$$M_\theta^* M_\theta + G^* G \leq I,$$

whence

$$\begin{aligned}\inf_{p \in \mathcal{P}(\mathcal{E}),\ p(0)=0} \langle G^*G(h-p), h-p\rangle &\leq \inf_{p \in \mathcal{P}(\mathcal{E}),\ p(0)=0} \langle (I - M_\theta^* M_\theta)(h-p), h-p\rangle \\ &= \langle \Delta(M_\theta)h, h\rangle = 0,\end{aligned}$$

for any $h \in \mathcal{E}$. Hence, and using Theorem 1.10, we obtain

$$0 = \inf_{p \in \mathcal{P}(\mathcal{E}),\ p(0)=0} \langle G^* G(h-p), h-p \rangle \geq \langle G(0)^* G(0) h, h \rangle$$

for any $h \in \mathcal{E}$. This implies $G(0) = 0$. Now, assume that $G \neq 0$ and let k be the smallest nonnegative integer such $G_{\alpha_0} \neq 0$ for some $\alpha_0 \in \mathbb{F}_n^+$ with $|\alpha_0| = k$. Using again Theorem 1.10, we get

$$0 = \inf_{p \in \mathcal{P}(\mathcal{E}),\ p(0)=0} \langle G^* G(h-p), h-p \rangle \geq \left\langle G_{\alpha_0}^* G_{\alpha_0} h, h \right\rangle$$

for any $h \in \mathcal{E}$, which implies $G_{\alpha_0} = 0$, a contradiction. Therefore, $G = 0$ and consequently M_θ is an extreme point of the unit ball of $R_n^\infty \bar{\otimes} B(\mathcal{E})$.

We prove now the second part of the theorem. Let $M_\theta \in R_n^\infty \bar{\otimes} B(\mathcal{E})$ and assume that $\dim \mathcal{E} < \infty$ and $\|M_\theta\| \leq 1$. Since the entropy of M_θ satisfies the equation $E(M_\theta) = e(I - M_\theta^* M_\theta)$, we can apply Corollary 1.2 to the multi-Toeplitz operator $I - M_\theta^* M_\theta$ in order to find an outer multi-analytic operator $M_\psi \in R_n^\infty \bar{\otimes} B(\mathcal{E})$ such that

$$M_\theta^* M_\theta + M_\psi^* M_\psi \leq I. \tag{1.47}$$

Define the operators

$$X := M_\theta + M_\varphi M_\psi \quad \text{and} \quad Y := M_\theta - M_\varphi M_\psi.$$

Since the operators M_θ and M_φ have orthogonal ranges, we can assume that $\|M_\varphi\| \leq 1$. Using relation (1.47), we obtain

$$X^* X \leq M_\theta^* M_\theta + M_\psi^* M_\varphi^* M_\varphi M_\psi \leq 1.$$

Similarly, we get $\|Y\| \leq 1$. Since $M_\theta = \frac{1}{2}(X + Y)$ and $X \neq Y$, we deduce that M_θ is not an extreme point of the unit ball of $R_n^\infty \bar{\otimes} B(\mathcal{E})$. This completes the proof. □

Let us remark that if $\mathcal{E} = \mathbb{C}$, then $\Delta(M_\theta) \neq 0$ if and only if the entropy $E(M_\theta) > -\infty$.

Corollary 1.15. *If $\varphi \in F_n^\infty$ (resp. $\mathcal{A}_n$, the noncommutative disc algebra), $\|\phi\| \leq 1$, and $E(M_\varphi) = -\infty$, then φ is an extreme point of the unit ball of F_n^∞ (resp. $\mathcal{A}_n$).*

Whether or not the converse of this corollary is true remains an open problem. For the time being, according to Theorem 1.14, we have a weaker form, namely, if $\varphi \in F_n^\infty$, $\|\varphi\| \leq 1$, and there is $\psi \in F_n^\infty$ such that φ and ψ have orthogonal ranges, then φ is an extreme point of the unit ball of F_n^∞ if and only if the entropy $E(\varphi) = -\infty$.

In [**2**], we prove that $H^\infty(\mathbb{D})$ can be completely isometrically embedded into the noncommutative analytic Toeplitz algebra F_n^∞ by the mapping $f \mapsto f(S_1)$, $f \in H^\infty(\mathbb{D})$. Surprisingly, under this embedding, the extreme points of the unit ball of $H^\infty(\mathbb{D})$ remain so in the unit ball of F_n^∞.

Theorem 1.16. *A function $f \in H^\infty(\mathbb{D})$ is an extreme point of the unit ball of $H^\infty(\mathbb{D})$ if and only if $f(S_1)$ is an extreme point of the unit ball of F_n^∞.*

PROOF. If $f \in H^\infty(\mathbb{D})$ is an extreme point of the unit ball of $H^\infty(\mathbb{D})$, then according to [**31**],

$$E(f) = \frac{1}{2\pi}\int_{-\pi}^{\pi} \ln(1 - |f(e^{it})|^2)\, dt = -\infty.$$

Accorging to relation (1.46) and the remark that follows, it is clear that $E(f(S_1)) \leq E(f)$. Therefore, we have $E(f(S_1)) = -\infty$. From Corollary 1.15, it follows that $f(S_1)$ is an extreme point of the unit ball of F_n^∞.

Now, assume that $f \in H^\infty(\mathbb{D})$ is not an extreme point of the unit ball of $H^\infty(\mathbb{D})$. Then there exist $\varphi, \psi \in (H^\infty(\mathbb{D}))_1$, $\varphi \neq \psi$ such that $f = \frac{\varphi+\psi}{2}$. Since the map $f \mapsto f(S_1)$ is a complete isometry of $H^\infty(\mathbb{D})$ into F_n^∞, and $f(S_1) = \frac{\varphi(S_1)+\psi(S_1)}{2}$, we deduce that $f(S_1)$ is not an extreme point of the unit ball of F_n^∞. This completes the proof. □

This result shows that there are extreme points of the unit ball $(F_n^\infty)_1$ which are not inner operators. Indeed, if $f \in H^\infty(\mathbb{D})$ is an extreme point of the unit ball of $H^\infty(\mathbb{D})$ such that f is not an inner function (see [**31**]), then, according to Theorem 1.16, $f(S_1)$ and $S_i f(S_1)$, $i = 2, \ldots, n$, are also extreme points of the unit ball of F_n^∞ which are not inner operators. For some other examples, take $\varphi(S_2, \ldots, S_n) \in F_n^\infty$ to be inner operator such that $\varphi(0) = 0$, and note that $\varphi(S_2, \ldots, S_n) f(S_1)$ has the required property.

CHAPTER 2

Noncommutative Commutant Lifting Theorem: Geometric Structure and Maximal Entropy Solution

We obtain several geometric characterizations for the multivariable central intertwining lifting, a maximum principle, and a permanence principle for the noncommutative commutant lifting theorem. Under certain natural conditions, we find explicit forms for the maximal entropy solution of this multivariable commutant lifting theorem, and concrete formulas for its entropy.

2.1. Multivariable intertwining liftings and geometric structure

In this section, we obtain some results concerning the geometric structure of the intertwining liftings in the noncommutative commutant lifting theorem (see [**38**] and [**41**]). It is shown that there is a one-to-one correspondence between the set of all intertwining liftings with tolerance $t > 0$ and certain families of contractions $\{C_k\}_{k=1}^{\infty}$ and $\{\Lambda_\alpha\}_{\alpha\in\mathbb{F}_n^+}$ (see Theorem 2.2 and Theorem 2.3). The geometric structure of the multivariable intertwining liftings will play an important role in our investigation.

Let us recall from [**37**], [**38**], and [**39**] a few results concerning the noncommutative dilation theory for sequences of operators (see [**66**] for the classical case $n = 1$). A sequence of operators $\mathcal{T} := (T_1, \ldots, T_n)$, $T_i \in B(\mathcal{H})$, is called row contraction if

$$T_1T_1^* + \cdots + T_nT_n^* \leq I_{\mathcal{H}}.$$

We say that a sequence of isometries $\mathcal{V} := (V_1, \ldots, V_n)$, $V_i \in B(\mathcal{K})$, is a minimal isometric dilation (m.i.d.) of $\mathcal{T}$ on a Hilbert space $\mathcal{K} \supseteq \mathcal{H}$ if the following properties are satisfied:

(i) $V_i^*V_j = 0$ for all $i \neq j$, $i, j \in \{1, \ldots, n\}$;
(ii) $V_j^*|\mathcal{H} = T_j^*$ for all $j = 1, \ldots, n$;
(iii) $\mathcal{K} = \bigvee_{\alpha\in\mathbb{F}_n^+} V_\alpha\mathcal{H}$.

If $\mathcal{V}$ satisfies only the condition (i) and $P_{\mathcal{H}}V_i = T_iP_{\mathcal{H}}$, $i = 1, \ldots, n$, then $\mathcal{V}$ is called isometric lifting of $\mathcal{T}$. The minimal isometric dilation of $\mathcal{T}$ is an isometric lifting and is uniquely determined up to an isomorphism [**38**].

Let us consider a canonical realization of it on Fock spaces. For convenience of notation, we will sometimes identify the n-tuple $(T_1, \ldots, T_n)$ with the row operator $[T_1 \ \cdots \ T_n]$. Define the defect operator $D_{\mathcal{T}} : \oplus_{j=1}^n \mathcal{H} \to \oplus_{j=1}^n \mathcal{H}$ by setting $D_{\mathcal{T}} := (I_{\oplus_{j=1}^n \mathcal{H}} - \mathcal{T}^*\mathcal{T})^{1/2}$, and let $\mathcal{D} := \overline{D_{\mathcal{T}}(\oplus_{j=1}^n \mathcal{H})}$, where $\otimes_{j=1}^n \mathcal{H}$ denotes the direct sum of n copies of $\mathcal{H}$. Let $D_i:\mathcal{H}\to 1\otimes\mathcal{D}\subset F^2(H_n)\otimes\mathcal{D}$ be defined by

$$D_ih := 1\otimes D_{\mathcal{T}}(\underbrace{0,\dots,0}_{i-1\text{ times}},h,0,\dots).$$

Consider the Hilbert space $\mathcal{K}:=\mathcal{H}\oplus[F^2(H_n)\otimes\mathcal{D}]$ and define $V_i:\mathcal{K}\to\mathcal{K}$ by

$$V_i(h\oplus\xi):=T_ih\oplus[D_ih+(S_i\otimes I_{\mathcal{D}})\xi] \tag{2.1}$$

for any $h\in\mathcal{H},\ \xi\in F^2(H_n)\otimes\mathcal{D}$. Note that

$$V_i=\begin{bmatrix}T_i & 0\\ D_i & S_i\otimes I_{\mathcal{D}}\end{bmatrix} \tag{2.2}$$

with respect to the decomposition $\mathcal{K}=\mathcal{H}\oplus[F^2(H_n)\otimes\mathcal{D}]$. In [**38**], we proved that $\mathcal{V}:=[V_1\ \cdots\ V_n]$ is the minimal isometric dilation of $\mathcal{T}$. Let $\mathcal{H}_0:=\mathcal{H}$ and

$$\mathcal{H}_k:=\mathcal{H}_{k-1}\bigvee\Big(\bigvee_{|\alpha|=1}V_\alpha\mathcal{H}_{k-1}\Big)\quad\text{if } k\geq 2.$$

Note that $\mathcal{K}=\bigvee_{k=0}^\infty\mathcal{H}_k,\ \ \mathcal{H}_k\subset\mathcal{H}_{k+1}$, and all subspaces $\mathcal{H}_k$ are invariant under each $V_i^*,\ i=1,\dots,n$. On the other hand, we have $\mathcal{H}_1=\mathcal{H}\oplus\mathcal{D}$ and

$$\mathcal{H}_k=\mathcal{H}\oplus\bigoplus_{|\alpha|\leq k-1}e_\alpha\otimes\mathcal{D}\quad\text{if } k\geq 2.$$

Denote $\mathcal{V}_0:=\mathcal{T}$ and $\mathcal{V}_k:=[V_{1,k}\ \cdots\ V_{n,k}]$ if $k\geq 1$, where $V_{i,k}:=P_{\mathcal{H}_k}V_i|_{\mathcal{H}_k}$, for any $i=1,\dots,n$, and $P_{\mathcal{H}_k}$ is the orthogonal projection from $\mathcal{K}$ onto $\mathcal{H}_k$. Note that the operators $V_{i,k},\ i=1,\dots,n$, are partial isometries with orthogonal final spaces and initial space $\mathcal{H}_{k-1}$. It is easy to see that $\mathcal{V}$ is also the minimal isometric dilation of $\mathcal{V}_k$, $V_i^*|_{\mathcal{H}_k}=V_{i,k}^*$, and

$$V_i^*=\text{SOT}-\lim_{k\to\infty}V_{i,k}^*P_{\mathcal{H}_k}$$

for any $i=1,\dots,n$. On the other hand, let us mention that $\mathcal{V}_{k+1}$ is the one-step dilation of $\mathcal{V}_k$, i.e., $\mathcal{H}_{k+1}=\mathcal{H}_k\oplus D_{\mathcal{V}_k}(\oplus_{j=1}^n\mathcal{H}_k)$, and, for each $i=1,\dots,n$,

$$V_{i,k+1}(x\oplus y)=V_{i,k}x\oplus D_{\mathcal{V}_k}(\underbrace{0,\dots,0}_{i-1\text{ times}},x,0,\dots)$$

for any $x\in\mathcal{H}_k$ and $y\in D_{\mathcal{V}_k}(\oplus_{j=1}^n\mathcal{H}_k)$.

Let $\mathcal{T}':=[T_1'\ \cdots\ T_n']$, $T_i'\in B(\mathcal{H}')$, be another row contraction and let $\mathcal{V}':=[V_1'\ \cdots\ V_n']$ be the minimal isometric dilation of $\mathcal{T}'$ on the Hilbert space $\mathcal{K}':=\mathcal{H}'\oplus[F^2(H_n)\otimes\mathcal{D}_{\mathcal{T}'}]$. Let $A\in B(\mathcal{H},\mathcal{H}')$ be an operator satisfying $AT_i=T_i'A$, for any $i=1,\dots,n$.

An intertwining lifting of A is an operator $B\in B(\mathcal{K},\mathcal{K}')$ satisfying $BV_i=V_i'B$, for any $i=1,\dots,n$, and $P_{\mathcal{H}'}B=AP_{\mathcal{H}}$. Let $\mathcal{V}_k':=[V_{1,k}'\ \cdots\ V_{n,k}']$, $k\geq 0$, $V_{i,k}':=P_{\mathcal{H}_k}V_i'|\mathcal{H}_k$, be the k-step dilation of $\mathcal{T}'$. An operator $A_k\in B(\mathcal{H}_k,\mathcal{H}_k')$ is a k-step intertwining lifting of $A_0:=A$, if

$$V_{i,k}'A_k=A_kV_{i,k}\quad i=1,\dots,n,$$

and $P_{\mathcal{H}'}A_k=AP_{\mathcal{H}}|\mathcal{H}_k$. A sequence $\{A_k\}_{k=0}^\infty$ of k-step intertwining liftings of A is compatible if

$$P_{\mathcal{H}_k'}A_{k+1}=A_kP_{\mathcal{H}_k}|\mathcal{H}_{k+1},\quad k=0,1,\dots.$$

Note that the sequence $\{A_k\}_{k=0}^\infty$ is compatible if A_{k+1} is a one-step intertwining lifting of A_k for any $k=0,1,\dots$. According to [**41**] (see [**21**] for the classical case

$n = 1$), there is a one-to-one correspodence between the set of all multivariable intertwining liftings B of A such that $\|B\| \leq 1$ and the set of all compatible sequences $\{A_k\}_{k=0}^{\infty}$, $\|A_k\| \leq 1$, of k-step intertwining liftings of A. Moreover the correspondence is given by

$$B = \text{SOT} - \lim_{k\to\infty} A_k P_{\mathcal{H}_k} \quad \text{and} \quad A_k = P_{\mathcal{H}_k} B|\mathcal{H}_k, \quad k = 0, 1, \ldots.$$

The noncommutative commutant lifting theorem (see [**38**], [**40**], [**41**]) states that there always exists an intertwining lifting B of A with $\|B\| = \|A\|$ (see [**66**], [**21**] for the classical case).

Define the subspaces

$$\begin{aligned} \mathcal{N} &:= \left\{ \sum_{i=1}^n D_A T_i h_i \oplus D_T(\oplus_{i=1}^n h_i) : \ \oplus_{i=1}^n h_i \in \oplus_{i=1}^n \mathcal{H} \right\}^{-}, \\ \mathcal{M} &:= [\mathcal{D}_A \oplus \mathcal{D}_T] \ominus \mathcal{N}. \end{aligned} \tag{2.3}$$

and the operator $W : \mathcal{N} \to \mathcal{D}_{T'} \oplus (\oplus_{i=1}^n \mathcal{D}_A)$ by setting

$$W\left(\sum_{i=1}^n D_A T_i h_i \oplus D_T(\oplus_{i=1}^n h_i \right) := D_{T'}(\oplus_{i=1}^n A h_i) \oplus (\oplus_{i=1}^n D_A h_i). \tag{2.4}$$

Since $AT_i = T_i' A$, $i = 1, \ldots, n$, we have

$$\begin{aligned} &\left\| \sum_{i=1}^n D_A T_i h_i \right\|^2 + \|D_T(\oplus_{i=1}^n h_i)\|^2 \\ &\qquad = \left\| \sum_{i=1}^n T_i h_i \right\|^2 - \left\| \sum_{i=1}^n A T_i h_i \right\|^2 + \|D_T(\oplus_{i=1}^n h_i)\|^2 \\ &\qquad = \| \oplus_{i=1}^n h_i\|^2 - \left\| \sum_{i=1}^n A T_i h_i \right\|^2 \\ &\qquad = \sum_{i=1}^n (\|D_A h_i\|^2 + \|A h_i\|^2) - \left\| \sum_{i=1}^n T_i' A h_i \right\|^2 \\ &\qquad = \sum_{i=1}^n \|D_A h_i\|^2 + \|D_{T'}(\oplus_{i=1}^n A h_i)\|^2. \end{aligned}$$

Due to these calculations, it is clear that W is an isometry. Consider the subspaces

$$\begin{aligned} \mathcal{N}' &:= \left\{ \sum_{i=1}^n D_{T'}(\oplus_{i=1}^n A h_i) \oplus (\oplus_{i=1}^n D_A h_i) : \ \oplus_{i=1}^n h_i \in \oplus_{i=1}^n \mathcal{H} \right\}^{-}, \\ \mathcal{M}' &:= [\mathcal{D}_{T'} \oplus (\oplus_{i=1}^n \mathcal{D}_A)] \ominus \mathcal{N}', \end{aligned}$$

and note that $W : \mathcal{N} \to \mathcal{N}'$ is a unitary operator. As in [**41**], one can prove that

$$\overline{P_{\mathcal{M}'}(\mathcal{D}_{T'} \oplus \{0\})} = \mathcal{M}'. \tag{2.5}$$

Consider the orthogonal projections:

$$\begin{aligned} P_{\mathcal{D}'} &: \mathcal{D}_{T'} \oplus (\oplus_{j=1}^n \mathcal{D}_A) \to \mathcal{D}_{T'}, \qquad P_{\mathcal{D}'}[d \oplus (\oplus_{j=1}^n d_j)] := d, \\ P_i &: \mathcal{D}_{T'} \oplus (\oplus_{j=1}^n \mathcal{D}_A) \to \mathcal{D}_A, \qquad P_i[d \oplus (\oplus_{j=1}^n d_j)] := d_i \end{aligned}$$

for any $i = 1, \ldots, n$.

Taking into account the results from [**41**] (see [**21**] for the case $n = 1$), we can prove the following.

THEOREM 2.1. *Let $\mathcal{T} := [T_1 \ \cdots \ T_n]$, $T_i \in B(\mathcal{H})$, and $\mathcal{T}' = [T_1' \ \cdots \ T_n']$, $T_i' \in B(\mathcal{H}')$, be row contractions, and let $\mathcal{V}$ and $\mathcal{V}'$ be the corresponding minimal isometric dilations, respectively. Let $A \in B(\mathcal{H}, \mathcal{H}')$ be a contraction such that $AT_i = T_i' A$ for any $i = 1, \dots, n$. Then there is a one-to-one correspondence between the set of all contractive one-step intertwining liftings A_1 of A and the set of all contractions $C : \mathcal{M} \to \mathcal{M}'$, and any contractive one-step intertwining lifting $A_1 : \mathcal{H} \oplus \mathcal{D}_\mathcal{T} \to \mathcal{H}' \oplus \mathcal{D}_{\mathcal{T}'}$ of A is given by*

$$A_1 = \begin{bmatrix} I_\mathcal{H} & 0 \\ 0 & P_{\mathcal{D}'}(WP_\mathcal{N} + CP_\mathcal{M}) \end{bmatrix} \begin{bmatrix} AP_\mathcal{H} \\ D_A \oplus I_{\mathcal{D}_\mathcal{T}} \end{bmatrix}, \tag{2.6}$$

where $C : \mathcal{M} \to \mathcal{M}'$ is a contraction.

Moreover, there exists a unitary operator $\Omega : \mathcal{D}_{A_1} \to (\oplus_{j=1}^n \mathcal{D}_A) \oplus \mathcal{D}_C$ satisfying

$$\Omega D_{A_1} = \begin{bmatrix} \begin{bmatrix} P_1 \\ \vdots \\ P_n \end{bmatrix} (WP_\mathcal{N} + CP_\mathcal{M}) \\ D_C P_\mathcal{M} \end{bmatrix} (D_A \oplus I_\mathcal{D}). \tag{2.7}$$

PROOF. Let $A_1 : \mathcal{H} \oplus \mathcal{D}_\mathcal{T} \to \mathcal{H}' \oplus \mathcal{D}_{\mathcal{T}'}$ be a contractive one-step intertwining lifting of A. Since $\|A_1\| \leq 1$, $A_1^*|\mathcal{H}' = A^*$, and taking into account the geometric structure of 2×2 matrix contractions, we deduce that

$$A_1 = \begin{bmatrix} A & 0 \\ XD_A & Y \end{bmatrix}, \tag{2.8}$$

where $[X \quad Y] : \mathcal{D}_A \oplus \mathcal{D}_\mathcal{T} \to \mathcal{D}_{\mathcal{T}'}$ is a contraction. One the other hand, we must have

$$\begin{bmatrix} A & 0 \\ XD_A & Y \end{bmatrix} \begin{bmatrix} T_i & 0 \\ D_i & 0 \end{bmatrix} = \begin{bmatrix} T_i & 0 \\ D_i & 0 \end{bmatrix} \begin{bmatrix} A & 0 \\ XD_A & Y \end{bmatrix}$$

for any $i = 1, \dots, n$. This equality holds if and only if

$$AT_i = T_i' A \quad \text{and} \quad XD_A T_i + YD_i = D_i A$$

for any $i = 1, \dots, n$. The latter relation is equivalent to

$$[X \quad Y] \begin{bmatrix} D_A T_i \\ D_i \end{bmatrix} = D_i A, \quad i = 1, \dots, n,$$

which shows that $[X \quad Y]|\mathcal{N} = F_0$, where the operator $F_0 : \mathcal{N} \to \mathcal{D}_{\mathcal{T}'}$ is defined by

$$F_0 \left[\sum_{i=1}^n D_A T_i h_i \oplus D_\mathcal{T}(\oplus_{i=1}^n h_i) \right] := D_{\mathcal{T}'}(\oplus_{i=1}^n A h_i). \tag{2.9}$$

It is well known ([**21**]) that the set of contractions satisfying the operator equation $[X \quad Y]|\mathcal{N} = F_0$ is given by

$$[X \quad Y] = F_0 P_\mathcal{N} + D_{F_0^*} F_1 P_\mathcal{M},$$

where $F_1 : \mathcal{M} \to \mathcal{D}_{F_0^*}$ is an arbitrary contraction. Then any contractive one-step intertwining lifting A_1 of A is given by

$$A_1 = \begin{bmatrix} I_\mathcal{H} & 0 \\ 0 & F_0 P_\mathcal{N} + D_{F_0^*} F_1 P_\mathcal{M} \end{bmatrix} \begin{bmatrix} AP_\mathcal{H} \\ D_A \oplus I_{\mathcal{D}_\mathcal{T}} \end{bmatrix}, \tag{2.10}$$

where $F_1 : \mathcal{M} \to \mathcal{D}_{F_0^*}$ is an arbitrary contraction. Moreover, there is a one-to-one correspondence between the set of all contractive one-step intertwining liftings A_1 of A and the set of all contractions $F_1 : \mathcal{M} \to \mathcal{D}_{F_0^*}$.

Since

$$F_0 = P_{\mathcal{D}'} W, \tag{2.11}$$

we have

$$\begin{aligned}\|D_{F_0^*} h\|^2 &= \|h\|^2 - \|F_0^*\|^2 = \|h\|^2 - \|W^* P_{\mathcal{N}'} P_{\mathcal{D}'}^* h\|^2 \\ &= \|h\|^2 - \|P_{\mathcal{N}'} P_{\mathcal{D}'}^* h\|^2 = \|P_{\mathcal{M}'} P_{\mathcal{D}'}^* h\|^2\end{aligned}$$

for any $h \in \mathcal{D}'$. Hence, the relation $RD_{F_0^*} = P_{\mathcal{M}'} P_{\mathcal{D}'}^*$ defines an isometry from $\mathcal{D}_{F_0^*}$ into $\mathcal{M}'$. Using relation (2.5), we infer that R is onto. Therefore, the operator $R^* : \mathcal{M}' \to \mathcal{D}_{F_0^*}$ is unitary and

$$D_{F_0^*} R^* = P_{\mathcal{D}'} | \mathcal{M}'. \tag{2.12}$$

Now, using relations (2.10), (2.11), and (2.12), we get

$$\begin{aligned} A_1 &= \begin{bmatrix} I_{\mathcal{H}} & 0 \\ 0 & P_{\mathcal{D}'}(W P_{\mathcal{N}} + D_{F_0^*} R^* R F_1 P_{\mathcal{M}}) \end{bmatrix} \begin{bmatrix} A P_{\mathcal{H}} \\ D_A \oplus I_{\mathcal{D}_T} \end{bmatrix} \\ &= \begin{bmatrix} I_{\mathcal{H}} & 0 \\ 0 & P_{\mathcal{D}'}(W P_{\mathcal{N}} + (P_{\mathcal{D}'}|\mathcal{M}') R F_1 P_{\mathcal{M}}) \end{bmatrix} \begin{bmatrix} A P_{\mathcal{H}} \\ D_A \oplus I_{\mathcal{D}_T} \end{bmatrix}, \end{aligned}$$

which proves relation (2.6), where $C := RF_1 : \mathcal{M} \to \mathcal{M}'$. Now, define the operator

$$V : \mathcal{H} \oplus \mathcal{D} \to \mathcal{H}' \oplus \left[\mathcal{D}' \oplus \left(\bigoplus_{i=1}^n \mathcal{D}_A\right)\right] \oplus \mathcal{D}_C \simeq (\mathcal{H}' \oplus \mathcal{D}') \oplus \left(\bigoplus_{i=1}^n \mathcal{D}_A \oplus \mathcal{D}_C\right),$$

by setting

$$V := \begin{bmatrix} I & 0 \\ 0 & W P_{\mathcal{N}} + C P_{\mathcal{M}} \\ 0 & D_C P_{\mathcal{M}} \end{bmatrix} \begin{bmatrix} A P_{\mathcal{H}} \\ D_A \oplus I_{\mathcal{D}} \end{bmatrix}.$$

Note that V is a product of two isometries. According to relation (2.6), we have $A_1 = P_{\mathcal{H}' \oplus \mathcal{D}'} V$. For each $x \in \mathcal{H} \oplus \mathcal{D}$, we get

$$\begin{aligned}\|D_{A_1} x\|^2 &= \|x\|^2 - \|A_1 x\|^2 = \|Vx\|^2 - \|A_1 x\|^2 \\ &= \left\| P_{\left(\bigoplus_{i=1}^n \mathcal{D}_A\right) \oplus \mathcal{D}_C} V x \right\|^2.\end{aligned}$$

Hence, it is clear that Ω defines an isometry from $\mathcal{D}_{A_1}$ to $(\bigoplus_{i=1}^n \mathcal{D}_A) \oplus \mathcal{D}_C$. On the other hand, for any $h_i \in \mathcal{H}$, we have

$$\begin{aligned}\Omega D_{A_1} \left(\sum_{i=1}^n T_i h_i \oplus D_T(\oplus_{i=1}^n h_i) \right) &= \begin{bmatrix} P_1 \\ \vdots \\ P_n \end{bmatrix} W \left(\sum_{i=1}^n D_A T_i h_i \oplus D_T(\oplus_{i=1}^n h_i) \right) \oplus 0 \\ &= \bigoplus_{i=1}^n D_A h_i \oplus 0.\end{aligned}$$

Hence, the range of Ω contains the subspace $\bigoplus_{i=1}^n D_A h_i \oplus \{0\}$. Notice that, if $x \in \mathcal{M}$, then $\Omega D_{A_1} y = z \oplus D_C x$ for some $z \in \bigoplus_{i=1}^n \mathcal{D}_A$. It is clear that $\{0\} \oplus \mathcal{D}_C \mathcal{M}$ is also in the range of Ω. Therefore, Ω is unitary. The proof is complete. □

A repeated application of Theorem 2.1 provides compatible contractive sequences $\{A_k\}_{k=0}^\infty$ of k-step intertwining liftings of A. Setting $B := \text{SOT}-\lim_{k\to\infty} A_k P_{\mathcal{H}_k}$, we get an intertwining lifting B of A satisfying $\|A\| = \|B\|$, which proves the noncommutative commutant lifting theorem. For details, see [**41**] (resp. [**21**] for the classical case $n = 1$).

We should mention that, as in the classical case, the general setting of the noncommutative commutant lifting theorem can be reduced to the case when $\mathcal{T} := [T_1 \cdots T_n]$ is a row isometry (see Corollary 2.2 in [**55**]).

Due to this reason, we will assume from now on that $\mathcal{T} := [T_1 \cdots T_n]$ is an isometry. Let $\mathcal{V}_k' := [V_{1,k}' \cdots V_{n,k}']$, $k \geq 0$, $V_{i,k}' := P_{\mathcal{H}_k} V_i'|\mathcal{H}_k$, be the k-step dilation of $\mathcal{T}' := [T_1' \cdots T_n']$. Let $A \in B(\mathcal{H}, \mathcal{H}')$, $t > 0$, and assume that $\|A\| \leq t$. Define the defect operator $D_{A,t} := (t^2 I - A^*A)^{1/2}$ and the subspace $\mathcal{D}_{A,t} := \overline{D_{A,t}\mathcal{H}}$. In the particular case when $t = 1$, we use the classical notation $D_{A,1} := D_A$ and $\mathcal{D}_{A,1} := \mathcal{D}_A$.

We say that $A_k : \mathcal{H} \to \mathcal{H}_k'$ is a k-step intertwining lifting of $A_0 := A$ with tolerance $t > 0$ if $\|A_k\| \leq t$, $V_{i,k}' A_k = A_k T_i$, for any $i = 1, \ldots, n$, and $P_{\mathcal{H}'} A_k = A$. Note that $\mathcal{D}_k' := \mathcal{D}_{\mathcal{V}_k'}$ can be identified with $\bigoplus_{|\alpha|=k} e_\alpha \otimes \mathcal{D}'$, where $\mathcal{D}' := \mathcal{D}_{\mathcal{T}'}$. For each $k \geq 0$, define the following subspaces:

$$\begin{aligned}
\mathcal{N}_k &:= \left\{ \sum_{i=1}^n D_{A_k,t} T_i h_i : \oplus_{i=1}^n h_i \in \oplus_{i=1}^n \mathcal{H} \right\}^- \subseteq \mathcal{D}_{A_k,t}, \\
\mathcal{M}_k &:= \mathcal{D}_{A_k,t} \ominus \mathcal{N}_k \subseteq \mathcal{D}_{A_k,t},
\end{aligned} \tag{2.13}$$

and

$$\begin{aligned}
\mathcal{N}_k' &:= \left\{ \sum_{i=1}^n D_k'(\oplus_{i=1}^n A_k h_i) \oplus (\oplus_{i=1}^n D_{A_k,t} h_i) : \oplus_{i=1}^n h_i \in \oplus_{i=1}^n \mathcal{H} \right\}^-, \\
\mathcal{M}_k' &:= [\mathcal{D}_k' \oplus (\oplus_{i=1}^n \mathcal{D}_{A_k,t})] \ominus \mathcal{N}_k',
\end{aligned}$$

where $D_k' := D_{\mathcal{V}_k'}$. Note that $\mathcal{N}_0 = \mathcal{N}$, $\mathcal{M}_0 = \mathcal{M}$, $\mathcal{N}_0' = \mathcal{N}'$, and $\mathcal{M}_0' = \mathcal{M}'$. Consider the orthogonal projections:

$$\begin{aligned}
&P_{\mathcal{D}_k'} : \mathcal{D}_k' \oplus (\oplus_{j=1}^n \mathcal{D}_{A_k,t}) \to \mathcal{D}_k', \qquad && P_{\mathcal{D}_k'}[d \oplus (\oplus_{j=1}^n d_j)] := d, \\
&P_{i,k} : \mathcal{D}_k' \oplus (\oplus_{j=1}^n \mathcal{D}_{A_k,t}) \to \mathcal{D}_{A_k,t}, \qquad && P_{i,k}[d \oplus (\oplus_{j=1}^n d_j)] := d_i
\end{aligned}$$

for any $i = 1, \ldots, n$. Note that $P_{\mathcal{D}_0'} = P_{\mathcal{D}'}$ and $P_{i,0} = P_i$ for any $i = 1, \ldots, n$. Define the unitary operator $W_k : \mathcal{N}_k \to \mathcal{N}_k'$ by

$$W_k \left(\sum_{i=1}^n D_{A_k,t} T_i h_i \right) := D_k'(\oplus_{i=1}^n A_k h_i) \oplus (\oplus_{i=1}^n D_{A_k,t} h_i). \tag{2.14}$$

Note that if $k = 0$ and $t = 1$, then we have $W_0 = W$ (see relation (2.4) in the particular case when $\mathcal{T}$ is a row isometry).

Now we can prove the following result.

THEOREM 2.2. *Let $\mathcal{T} := [T_1 \cdots T_n]$, $T_i \in B(\mathcal{H})$, be a row isometry and let $\mathcal{T}' := [T_1' \cdots T_n']$, $T_i' \in B(\mathcal{H}')$, be a row contraction with minimal isometric dilation $\mathcal{V}'$. Let $A \in B(\mathcal{H}, \mathcal{H}')$, $t > 0$, be such that $\|A\| \leq t$ and $AT_i = T_i' A$ for any $i = 1, \ldots, n$. For each $k = 0, 1, \ldots$, let A_k be a k-step intertwining lifting of A with*

$\|A_k\| \leq t$. *Then any one-step intertwining lifting* A_{k+1} *of* A_k *such that* $\|A_{k+1}\| \leq t$ *is given by*

$$A_{k+1} = \begin{bmatrix} A_k \\ P_{\mathcal{D}'_k}(W_k P_{\mathcal{N}_k} + C_k P_{\mathcal{M}_k}) D_{A_k,t} \end{bmatrix} : \mathcal{H} \to \mathcal{H}'_k \oplus \mathcal{D}'_k, \tag{2.15}$$

where $C_k : \mathcal{M}_k \to \mathcal{M}'_k$ *is a contraction. Moreover, there is a one-to-one correspondence between the set of all one-step intertwining liftings* A_{k+1} *of* A_k *with* $\|A_{k+1}\| \leq t$ *and the set of all contractions* $C_k : \mathcal{M}_k \to \mathcal{M}'_k$, *and there exists a unitary operator* $\Omega_k : \mathcal{D}_{A_{k+1},t} \to \left(\bigoplus_{j=1}^{n} \mathcal{D}_{A_k,t}\right) \oplus \mathcal{D}_{C_k}$ *satisfying*

$$\Omega_k D_{A_{k+1},t} = \begin{bmatrix} \begin{bmatrix} P_{1,k} \\ \vdots \\ P_{n,k} \end{bmatrix} (W_k P_{\mathcal{N}_k} + C_k P_{\mathcal{M}_k}) D_{A_k,t} \\ D_{C_k} P_{\mathcal{M}_k} D_{A_k,t} \end{bmatrix}. \tag{2.16}$$

PROOF. Let $A \in B(\mathcal{H}, \mathcal{H}')$ be such that $\|A\| \leq t$ and $AT_i = T'_i A$ for any $i = 1, \ldots, n$. Note that since $\mathcal{T}$ is a row isometry and $\|A\| \leq t$, Theorem 2.1 remains true in a slightly adapted version. More precisely, taking into account that $\mathcal{D}_{\mathcal{T}} = \{0\}$ and replacing D_A with $D_{A,t}$, we deduce that any one-step intertwining lifting A_1 of A such that $\|A_1\| \leq t$ is given by

$$A_1 = \begin{bmatrix} A \\ P_{\mathcal{D}'}(W P_{\mathcal{N}} + C P_{\mathcal{M}}) D_{A,t} \end{bmatrix} : \mathcal{H} \to \mathcal{H}' \oplus \mathcal{D}_{\mathcal{T}'}. \tag{2.17}$$

where $C : \mathcal{M} \to \mathcal{M}'$ is a contraction. Moreover, there is a one-to-one correspondence between the set of all one-step intertwining liftings A_1 of A such that $A_1\| \leq t$ and the set of all contractions $C : \mathcal{M} \to \mathcal{M}'$, and there exists a unitary operator $\Omega : \mathcal{D}_{A_1,t} \to (\oplus_{j=1}^{n} \mathcal{D}_{A,t}) \oplus \mathcal{D}_C$ satisfying

$$\Omega D_{A_1,t} = \begin{bmatrix} \begin{bmatrix} P_1 \\ \vdots \\ P_n \end{bmatrix} (W P_{\mathcal{N}} + C P_{\mathcal{M}}) D_{A,t} \\ D_C P_{\mathcal{M}} D_{A,t} \end{bmatrix}.$$

Since A_{k+1} is a one-step intertwining lifting of A_k, we can apply the first part of this proof to the k-step intertwining lifting A_k of A, and complete the proof. □

According to Theorem 2.2, there is a one-to-one correspondence between the set of all intertwining liftings of A with tolerance t, and the set of all contractions $C_k : \mathcal{M}_k \to \mathcal{M}'_k$, $k = 1, 2, \ldots$, given by

$$B = \text{SOT} - \lim_{k \to \infty} A_k P_{\mathcal{H}_k},$$

where the sequence $\{A_k\}$ is defined by (2.15). We remark that if $t = 1$, then this correspondence is, up to unitary operators, exactly the one obtained in [**41**] (see [**21**] for the case $n = 1$) between the set of all contractive intertwining liftings of A and the set of all generalized choice sequences.

Now, we show that there is a one-to-one correspondence between the set of all intertwining liftings with tolerance $t > 0$ and certain families of contractions $\{\Lambda_\alpha\}_{\alpha \in \mathbb{F}_n^+}$.

THEOREM 2.3. *Let $\mathcal{T} := [T_1 \ \cdots \ T_n]$, $T_i \in B(\mathcal{H})$, be a row isometry and let $\mathcal{T}' := [T_1' \ \cdots \ T_n']$, $T_i' \in B(\mathcal{H}')$, be a row contraction with its minimal isometric dilation $\mathcal{V}' := [V_1' \ \cdots \ V_n']$, $V_i' \in B(\mathcal{K}')$, on the Hilbert space $\mathcal{K}' := \mathcal{H}' \oplus [F^2(H_n) \otimes \mathcal{D}']$. Let $A \in B(\mathcal{H}, \mathcal{H}')$ be such that $\|A\| \leq t$ and*

$$AT_i = T_i'A, \quad i = 1, \dots, n.$$

Then B is an intertwining lifting of A with respect to $\mathcal{V}'$ and tolerance $t > 0$ if and only if B has a matrix decomposition

$$B = \begin{bmatrix} A \\ \Lambda D_{A,t} \end{bmatrix} : \mathcal{H} \to \mathcal{H}' \oplus [F^2(H_n) \otimes \mathcal{D}'], \tag{2.18}$$

where $\Lambda : \mathcal{D}_{A,t} \to F^2(H_n) \otimes \mathcal{D}'$ is a contraction with $\Lambda h = \sum_{\alpha \in \mathbb{F}_n^+} e_\alpha \otimes \Lambda_\alpha h$, $h \in \mathcal{D}_{A,t}$, and such that the operators $\Lambda_\alpha \in B(\mathcal{D}_{A,t}, \mathcal{D}')$ are given by

$$\begin{aligned} \Lambda_{g_0} &= P_{\mathcal{D}'} W P_{\mathcal{N}} + Y_{g_0} P_{\mathcal{M}} \\ \Lambda_{g_j \alpha} &= \Lambda_\alpha P_j W P_{\mathcal{N}} + Y_{g_j \alpha} P_{\mathcal{M}}, \end{aligned} \tag{2.19}$$

$j = 1, \dots, n$, for some contractions $Y_\alpha \in B(\mathcal{M}, \mathcal{D}')$, $\alpha \in \mathbb{F}_n^+$, where the subspaces $\mathcal{N}$ and $\mathcal{M}$ are defined by relation (2.13) (when $k = 0$). Moreover, the families of contractions $\{\Lambda_\alpha\}_{\alpha \in \mathbb{F}_n^+}$ and $\{Y_\alpha\}_{\alpha \in \mathbb{F}_n^+}$ uniquely determine each other.

PROOF. Since B is an intertwining lifting of A with respect lo $\mathcal{V}'$ we must have

$$B = \begin{bmatrix} A \\ X \end{bmatrix} : \mathcal{H} \to \mathcal{H}' \oplus [F^2(H_n) \otimes \mathcal{D}'].$$

It is easy to see that $\|B\| \leq t$ if and only if $\|Xh\| \leq \|D_{A,t}h\|$, and therefore, if and only if there is a contraction $\Lambda : \mathcal{D}_{A,t} \to F^2(H_n) \otimes \mathcal{D}'$ such that $X = \Lambda D_{A,t}$. Note that Λ is uniquely determined by B. Since $BT_i = V_i'B$ and $AT_i = T_i'A$ for any $i = 1, \dots, n$, we can use relation (2.2) and get

$$\Lambda D_{A,t} T_i = D_i' A + (S_i \otimes I_{\mathcal{D}'}) \Lambda D_{A,t} \tag{2.20}$$

for any $i = 1, \dots, n$. Setting

$$\Lambda h = \sum_{\alpha \in \mathbb{F}_n^+} e_\alpha \otimes \Lambda_\alpha h, \quad h \in \mathcal{D}_{A,t},$$

we infer that relation (2.20) is equivalent to

$$\sum_{\beta \in \mathbb{F}_n^+} e_\beta \otimes \Lambda_\beta D_{A,t} T_i h = 1 \otimes D_i' A h + \sum_{\alpha \in \mathbb{F}_n^+} e_{g_i \alpha} \otimes \Lambda_\alpha D_{A,t} h$$

for any $h \in \mathcal{D}_{A,t}$ and $i = 1, \dots, n$. It is clear that the latter equality is equivalent to the following equations:

$$\begin{aligned} \Lambda_{g_0} D_{A,t} T_i &= D_i' A \\ \Lambda_{g_j \alpha} D_{A,t} T_i &= \delta_{ij} \Lambda_\alpha D_{A,t} \end{aligned} \tag{2.21}$$

for any $i, j \in \{1, \dots, n\}$ and $\alpha \in \mathbb{F}_n^+$. We recall that

$$P_{\mathcal{M}} D_{A,t} T_i h = 0 \quad \text{and} \quad P_j W P_{\mathcal{N}} D_{A,t} T_i h = \delta_{ij} D_{A,t} h \tag{2.22}$$

for any $h \in \mathcal{H}$ and $i, j \in \{1, \dots, n\}$. On the other hand, relation (2.4) implies

$$P_{\mathcal{D}'} W \left(\sum_{i=1}^n D_{A,t} T_i h_i \right) = D'(\oplus_{i=1}^n A h_i). \tag{2.23}$$

Taking into account relations (2.22), (2.23), and (2.21), we deduce

$$\begin{aligned} P_{\mathcal{D}'}WP_{\mathcal{N}}D_{A,t}T_i &= P_{\mathcal{D}'}WD_{A,t}T_i = D_i'A \\ &= \Lambda_{g_0}D_{A,t}T_i = \Lambda_{g_0}P_{\mathcal{N}}D_{A,t}T_i \end{aligned}$$

for any $i = 1, \ldots, n$, which implies

$$\Lambda_{g_0} = P_{\mathcal{D}'}WP_{\mathcal{N}} + Y_{g_0}P_{\mathcal{M}},$$

where $Y_{g_0} \in B(\mathcal{M}, \mathcal{D}')$. Using again relations (2.22), (2.23), and (2.21), we obtain

$$\begin{aligned} \Lambda_\alpha P_j W P_{\mathcal{N}} D_{A,t} T_i &= \delta_{ij}\Lambda_\alpha D_{A,t} = \Lambda_{g_i\alpha} D_{A,t} T_i \\ &= \Lambda_{g_i\alpha} P_{\mathcal{N}} D_{A,t} T_i \end{aligned}$$

for any $i, j \in \{1, \ldots, n\}$, which implies

$$\Lambda_{g_j\alpha} = \Lambda_\alpha P_j W P_{\mathcal{N}} + Y_{g_j\alpha}P_{\mathcal{M}}, \quad j = 1, \ldots, n,$$

where $Y_{g_j\alpha} \in B(\mathcal{M}, \mathcal{D}')$. Now, it is easy to see that the families of contractions $\{\Lambda_\alpha\}_{\alpha\in\mathbb{F}_n^+}$ and $\{Y_\alpha\}_{\alpha\in\mathbb{F}_n^+}$ uniquely determine each other. The proof is complete. □

2.2. Central lifting in several variables and geometric characterizations

The main results of this section (see Theorem 2.4 and Theorem 2.5) show that the intertwining lifting corresponding to the parameters $C_k = 0$, $k = 1, 2, \ldots$, (resp. $\Lambda_\alpha = 0$, $\alpha \in \mathbb{F}_n^+$) coincides with the central intertwining lifting with tolerance t, for which we have an explicit form. At the end of this section, we show that, under certain conditions, there is only one intertwining lifting B of A such that $\|B\| = \|A\|$, namely the central intertwining lifting. The geometric structure of the central intertwining lifting will play a very important role in our investigation.

Our first result shows that the intertwining lifting of A corresponding to the parameters $C_k = 0$ for any $k = 1, 2, \ldots$, coincides with the central intertwining lifting B_c of A (as defined in [**53**] when $t = 1$).

THEOREM 2.4. *Let $A \in B(\mathcal{H}, \mathcal{H}')$, $t > 0$, be such that $\|A\| \leq t$ and $AT_i = T_i'A$ for any $i = 1, \ldots, n$, where $\mathcal{T} := [T_1 \cdots T_n]$, $T_i \in B(\mathcal{H})$, is an isometry and $\mathcal{T}' := [T_1' \cdots T_n']$, $T_i' \in B(\mathcal{H}')$, is a row contraction. Let $\mathcal{K}' := \mathcal{H}' \oplus [F^2(H_n) \otimes \mathcal{D}']$ and $\mathcal{V}' := [V_1' \cdots V_n']$, $V_i' \in B(\mathcal{K}')$, be the minimal isometric dilation of $\mathcal{T}'$.*

Then the intertwining lifting of A with tolerance t corresponding to the parameters $C_k = 0$, $k = 1, 2, \ldots$, is given by $B_c : \mathcal{H} \to \mathcal{H}' \oplus [F^2(H_n) \otimes \mathcal{D}']$ and

$$B_c h = Ah \oplus \sum_{\sigma\in\mathbb{F}_n^+} e_\sigma \otimes (P_{\mathcal{D}'}WP_{\mathcal{N}})E_{\tilde{\sigma}}D_{A,t}h, \quad h \in \mathcal{H}, \tag{2.24}$$

where $E_{g_0} := I_{\mathcal{D}_{A,t}}$ and $E_i := P_iWP_{\mathcal{N}}$ for any $i = 1, \ldots, n$. In particular, we have $B_c^*|\mathcal{H}' = A^*$ and $\|B_c\| \leq t$.

PROOF. Let $\{A_k\}_{k=0}^\infty$ be the set of the k-step intertwining liftings of A with $\|A_k\| \leq t$, obtained by setting $C_k = 0$ for any $k = 0, 1, \ldots$. Then, for each $k = 0, 1, \ldots$, the operator A_{k+1} is the one-step intertwining lifting of A_k when $C_k = 0$, and $A_0 = A$. According to Theorem 2.2, we have

$$A_1 h = Ah \oplus 1 \otimes (P_{\mathcal{D}'}WP_{\mathcal{N}})D_{A,t}h, \quad h \in \mathcal{H}.$$

By induction, we assume that

$$A_k h = Ah \oplus \sum_{|\sigma|\leq k-1} e_\sigma \otimes (P_{\mathcal{D}'} W P_{\mathcal{N}}) E_{\tilde{\sigma}} D_{A,t} h, \quad h \in \mathcal{H}. \tag{2.25}$$

Note that, for each $h \in \mathcal{H}$, we have

$$\begin{aligned}
t^2\|h\|^2 - \|A_k h\|^2 &= t^2\|h\|^2 - \|Ah\|^2 - \sum_{|\sigma|\leq k-1} \|(P_{\mathcal{D}'} W P_{\mathcal{N}}) E_{\tilde{\sigma}} D_{A,t} h\|^2 \\
&= \|D_{A,t}h\|^2 - \sum_{|\sigma|\leq k-1} \left(\|W P_{\mathcal{N}} E_{\tilde{\sigma}} D_{A,t} h\|^2 - \sum_{i=1}^n \|(P_i W P_{\mathcal{N}}) E_{\tilde{\sigma}} D_{A,t} h\|^2 \right) \\
&= \|D_{A,t}h\|^2 - \sum_{|\sigma|\leq k-1} \|P_{\mathcal{N}} E_{\tilde{\sigma}} D_{A,t} h\|^2 + \sum_{i=1}^n \sum_{|\sigma|\leq k-1} \|E_{\widetilde{\sigma g_i}} D_{A,t} h\|^2 \\
&= \|D_{A,t}h\|^2 - \|P_{\mathcal{N}} D_{A,t} h\|^2 - \sum_{1\leq|\sigma|\leq k-1} \|P_{\mathcal{N}} E_{\tilde{\sigma}} D_{A,t} h\|^2 + \sum_{1\leq|\omega|\leq k} \|E_{\tilde{\omega}} D_{A,t} h\|^2 \\
&= \|P_{\mathcal{M}} D_{A,t} h\|^2 + \sum_{1\leq|\sigma|\leq k-1} \|P_{\mathcal{M}} E_{\tilde{\sigma}} D_{A,t} h\|^2 + \sum_{|\omega|=k} \|E_{\tilde{\omega}} D_{A,t} h\|^2.
\end{aligned}$$

Hence, we infer that $\|A_k\| \leq t$ and

$$\|D_{A_k,t} h\|^2 = \sum_{|\sigma|\leq k-1} \|P_{\mathcal{M}} E_{\tilde{\sigma}} D_{A,t} h\|^2 + \sum_{|\omega|=k} \|E_{\tilde{\omega}} D_{A,t} h\|^2.$$

Now, it is clear that there is an isometry

$$Z_k : \mathcal{D}_{A_k,t} \to \left[\bigoplus_{|\sigma|\leq k-1} e_\sigma \otimes \mathcal{M} \right] \oplus \left[\bigoplus_{|\omega|=k} e_\omega \otimes \mathcal{D}_{A,t} \right]$$

satisfying the equation

$$Z_k D_{A_k,t} h = \left(\sum_{|\sigma|\leq k-1} e_\sigma \otimes P_{\mathcal{M}} E_{\tilde{\sigma}} D_{A,t} h \right) \oplus \left(\sum_{|\omega|=k} e_\omega \otimes E_{\tilde{\omega}} D_{A,t} h \right) \tag{2.26}$$

for any $h \in \mathcal{H}$.

We show now that Z_k is a unitary operator. First, note that

$$P_j W P_{\mathcal{N}} D_{A,t} T_i h = \delta_{i,j} D_{A,t} h \quad \text{and} \quad P_{\mathcal{M}} D_{A,t} T_i h = 0$$

for any $h \in \mathcal{H}$ and $i, j \in \{1, 2, \ldots, n\}$. Hence, we infer that if $|\sigma| = |\omega|$, then

$$E_{\tilde{\sigma}} D_{A,t} T_\omega h = \delta_{\omega,\sigma} D_{A,t} h, \quad h \in \mathcal{H}. \tag{2.27}$$

Hence, and using (2.26), we infer that if $|\omega| = k$, then

$$Z_k D_{A_k,t} \left(\sum_{|\omega|=k} T_\omega h_\omega \right) = \sum_{|\omega|=k} e_\omega \otimes D_{A,t} h_\omega.$$

On the other hand,

$$Z_k D_{A_k,t} \left(\sum_{|\omega|=k-1} T_\omega h_\omega \right) = \left(\sum_{|\omega|=k} e_\omega \otimes d_\omega \right) \oplus \left(\sum_{|\omega|=k-1} e_\omega \otimes P_{\mathcal{M}} D_{A,t} h_\omega \right)$$

for some $d_\omega \in \mathcal{D}_{A,t}$. The last two equations show that the closed linear span of the subspaces $Z_k D_{A_k,t}\left(\bigvee_{|\omega|=k-1} T_\omega \mathcal{H}\right)$ and $Z_k D_{A_k,t}\left(\bigvee_{|\omega|=k} T_\omega \mathcal{H}\right)$ coincides with

$$\left[\bigoplus_{|\omega|=k} e_\omega \otimes \mathcal{D}_{A,t}\right] \oplus \left[\bigoplus_{|\sigma|=k-1} e_\omega \otimes \mathcal{M}\right].$$

Continuing this process, one can show that Z_k is surjective, and consequently a unitary operator. Since $\mathcal{N}_k = \bigvee_{i=1}^n D_{A_k,t} T_i \mathcal{H}$, it is easy to see that

$$\begin{aligned} Z_k \mathcal{N}_k &= \left[\bigoplus_{|\omega|=k} e_\omega \otimes \mathcal{D}_{A,t}\right] \oplus \left[\bigoplus_{1\leq|\sigma|\leq k-1} e_\omega \otimes \mathcal{M}\right] \quad \text{and} \\ Z_k \mathcal{M}_k &= 1 \otimes \mathcal{M}. \end{aligned} \tag{2.28}$$

According to Theorem 2.2 (when $C_k = 0$), to complete our proof by induction, we need to show that

$$P_{\mathcal{D}'_k} W_k P_{\mathcal{N}_k} D_{A_k,t} h = \sum_{|\sigma|=k} e_\sigma \otimes (P_{\mathcal{D}'} W P_{\mathcal{N}}) E_{\tilde{\sigma}} D_{A,t} h, \quad h \in \mathcal{H}, \tag{2.29}$$

where $\mathcal{D}'_k$ is identified with $\bigoplus_{|\omega|=k} e_\omega \otimes \mathcal{D}'$ and $\mathcal{D}' := \mathcal{D}_{T'}$. Using relations (2.26) and (2.28), we obtain

$$\begin{aligned} &P_{\mathcal{D}'_k} W_k P_{\mathcal{N}_k} D_{A_k,t} h = P_{\mathcal{D}'_k} W_k P_{\mathcal{N}_k} Z_k^* Z_k D_{A_k,t} h \\ &\overset{(2.26)}{=} P_{\mathcal{D}'_k} W_k P_{\mathcal{N}_k} Z_k^* \left(\sum_{|\omega|=k} e_\omega \otimes E_{\tilde{\omega}} D_{A,t} h \oplus \sum_{|\sigma|\leq k-1} e_\sigma \otimes P_{\mathcal{M}} E_{\tilde{\sigma}} D_{A,t} h \right) \\ &\overset{(2.28)}{=} P_{\mathcal{D}'_k} W_k Z_k^* \left(\sum_{|\omega|=k} e_\omega \otimes E_{\tilde{\omega}} D_{A,t} h \oplus \sum_{1\leq|\sigma|\leq k-1} e_\sigma \otimes P_{\mathcal{M}} E_{\tilde{\sigma}} D_{A,t} h \right) \\ &= \sum_{j=1}^n P_{\mathcal{D}'_k} W_k Z_k^* \left(\sum_{|\alpha|=k-1} e_{g_j\alpha} \otimes E_{\tilde{\alpha}} E_{g_j} D_{A,t} h \oplus \sum_{0\leq|\beta|\leq k-2} e_{g_j\beta} \otimes P_{\mathcal{M}} E_{\tilde{\beta}} E_{g_j} D_{A,t} h \right). \end{aligned}$$

Now, using relations (2.27), (2.26), we deduce that

$$\begin{aligned} &P_{\mathcal{D}'_k} W_k Z_k^* \left(\sum_{|\alpha|=k-1} e_{g_j\alpha} \otimes E_{\tilde{\alpha}} D_{A,t} y \oplus \sum_{0\leq|\beta|\leq k-2} e_{g_j\beta} \otimes P_{\mathcal{M}} E_{\tilde{\beta}} D_{A,t} y \right) \\ &\overset{(2.27)}{=} P_{\mathcal{D}'_k} W_k Z_k^* \sum_{i=1}^n \left(\sum_{|\alpha|=k-1} e_{g_i\alpha} \otimes E_{g_i\tilde{\alpha}} D_{A,t} T_j y \oplus \sum_{0\leq|\beta|\leq k-2} e_{g_i\beta} \otimes P_{\mathcal{M}} E_{g_i\tilde{\beta}} D_{A,t} T_j y \right) \\ &\overset{(2.26)}{=} P_{\mathcal{D}'_k} W_k P_{\mathcal{N}_k} Z_k^* Z_k D_{A_k,t} T_j y = P_{\mathcal{D}'_k} W_k P_{\mathcal{N}_k} D_{A_k,t} T_j y \\ &\overset{(2.14)}{=} D'_k(\underbrace{0,\ldots,0}_{j-1 \text{ times}}, A_k y, 0, \ldots) = S_j P_{\bigoplus_{|\omega|=k}(e_\omega \otimes \mathcal{D}')} A_k y \\ &= \sum_{|\alpha|=k-1} e_{g_j} e_\alpha \otimes (P_{\mathcal{D}'} W P_{\mathcal{N}}) E_{\tilde{\alpha}} D_{A,t} y. \end{aligned}$$

Here, we used the fact that $D'_k = \overset{n}{\underset{j=1}{\oplus}} P_{\underset{|\omega|=k}{\oplus}(e_\omega \otimes \mathcal{D}')}$ and $\mathcal{D}'_k$ is identified with

$$\overset{n}{\underset{j=1}{\oplus}} \underset{|\omega|=k}{\oplus} (e_{g_j\omega} \otimes \mathcal{D}').$$

Now, since $D_{A,t}$ has dense range in $\mathcal{D}_{A,t}$, we infer that

$$\begin{aligned} P_{\mathcal{D}'_k} W_k Z_k^* \left(\sum_{|\alpha|=k-1} e_{g_j\alpha} \otimes E_{\tilde{\alpha}} x \oplus \sum_{0\leq|\beta|\leq k-2} e_{g_j\beta} \otimes P_{\mathcal{M}} E_{\tilde{\beta}} x \right) \\ = \sum_{|\alpha|=k-1} e_{g_j} e_\alpha \otimes (P_{\mathcal{D}'} W P_{\mathcal{N}}) E_{\tilde{\alpha}} x \end{aligned}$$

for any $x \in \mathcal{D}_{A,t}$ and $j = 1, 2, \ldots, n$. Summing up the results obtained so far, we deduce

$$\begin{aligned} P_{\mathcal{D}'_k} W_k P_{\mathcal{N}_k} D_{A_k,t} h &= \sum_{j=1}^n P_{\mathcal{D}'_k} W_k Z_k^* \left(\sum_{|\alpha|=k-1} e_{g_j\alpha} \otimes E_{\tilde{\alpha}} (E_{g_j} D_{A,t} h) \right. \\ &\qquad \left. \oplus \sum_{0\leq|\beta|\leq k-2} e_{g_j\beta} \otimes P_{\mathcal{M}} E_{\tilde{\beta}} (E_{g_j} D_{A,t} h) \right) \\ &= \sum_{j=1}^n \sum_{|\alpha|=k-1} e_{g_j\alpha} \otimes (P_{\mathcal{D}'} W P_{\mathcal{N}}) E_{\tilde{\alpha}} (E_{g_j} D_{A,t} h) \\ &= \sum_{|\beta|=k} e_\beta \otimes (P_{\mathcal{D}'} W P_{\mathcal{N}}) E_{\tilde{\beta}} D_{A,t} h). \end{aligned}$$

Therefore, relation (2.29) is proved. Since $\|A_k\| \leq k$ for any $k = 1, 2, \ldots$, we get $\|B_c\| \leq t$, and the proof is complete. □

Note that if $t = \|A\|$, then Theorem 2.4 implies that the central intertwining lifting B_c of A satisfies $\|B_c\| = \|A\|$.

The following result provides another characterization for the central intertwining lifting in the noncommutative commutant lifting theorem. We employ the notation from Theorem 2.3.

THEOREM 2.5. *Let $\mathcal{T} := [T_1 \cdots T_n]$, $T_i \in B(\mathcal{H})$, be a row isometry and let $\mathcal{T}' := [T'_1 \cdots T'_n]$, $T'_i \in B(\mathcal{H}')$, be a row contraction with its minimal isometric dilation $\mathcal{V}' := [V'_1 \cdots V'_n]$, $V'_i \in B(\mathcal{K}')$, on the Hilbert space $\mathcal{K}' := \mathcal{H}' \oplus [F^2(H_n) \otimes \mathcal{D}']$. Let $A \in B(\mathcal{H}, \mathcal{H}')$ be such that $\|A\| \leq t$ and*

$$AT_i = T'_i A, \quad i = 1, \ldots, n.$$

Let B be an intertwining liftig of A with tolerance $t > 0$, and let $B = \begin{bmatrix} A \\ \Lambda D_{A,t} \end{bmatrix}$ be the corresponding decomposition. Then the following statements are equivalent:

(i) *B is the central intertwining lifting of A with tolerance $t > 0$;*
(ii) *$\Lambda|\mathcal{M} = 0$;*
(iii) *$\Lambda_\alpha|\mathcal{M} = 0$ for any $\alpha \in \mathbb{F}_n^+$;*
(iv) *$Y_\alpha|\mathcal{M} = 0$ for any $\alpha \in \mathbb{F}_n^+$.*

In particular, if $\mathcal{M} = \{0\}$, then the central intertwining lifting of A with tolerance $t > 0$ is the unique intertwining liftig B of A such that $\|B\| \leq t$.

PROOF. Note that, using Theorem 2.3, Theorem 2.4, and relations (2.18), (2.19), and (2.24), the result follows. □

Let $\mathcal{T}' := [T_1' \ \cdots \ T_n']$, $T_i' \in B(\mathcal{H}')$, be a row contraction and let $\mathcal{V}' := [V_1' \ \cdots \ V_n']$, $V_i' \in B(\mathcal{K}')$, be its minimal isometric dilation on the Hilbert space $\mathcal{K}' := \mathcal{H}' \oplus [F^2(H_n) \otimes \mathcal{D}']$. If $\mathcal{W}' := [W_1' \ \cdots \ W_n']$, $W_i' \in B(\mathcal{G}')$, is an arbitrary isometric lifting of $\mathcal{T}'$ on a Hilbert space $\mathcal{G}' \supseteq \mathcal{H}'$, then there exists a unique isometry $\Phi : \mathcal{K}' \to \mathcal{G}'$ such that $\Phi V_i' = W_i' \Phi$, for any $i = 1, \ldots, n$, and $\Phi|\mathcal{H}' = I_{\mathcal{H}'}$. Moreover, using the geometric structure of the minimal isometric dilation (see [**38**]), one can deduce that Φ is given by

$$\Phi\left(h \oplus \sum_{\alpha \in \mathbb{F}_n^+} e_\alpha \otimes D_{\mathcal{T}'} h_\alpha\right) = h + \sum_{\alpha \in \mathbb{F}_n^+} W_\alpha' [W_1' - T_1' \ \cdots \ W_n' - T_n'] h_\alpha$$

for any $h \oplus \sum\limits_{\alpha \in \mathbb{F}_n^+} e_\alpha \otimes h_\alpha$ in $\mathcal{H}' \oplus [F^2(H_n) \otimes \mathcal{D}']$. Indeed, we have

$$\begin{aligned}
\Phi\left(0 \oplus \sum_{\alpha \in \mathbb{F}_n^+} e_\alpha \otimes D_{\mathcal{T}'} h_\alpha\right) &= \Phi\left(0 \oplus \sum_{\alpha \in \mathbb{F}_n^+} V_\alpha'(1 \otimes D_{\mathcal{T}'} h_\alpha)\right) \\
&= \sum_{\alpha \in \mathbb{F}_n^+} \Phi V_\alpha' [V_1' - T_1' \ \cdots \ V_n' - T_n'] h_\alpha \\
&= \sum_{\alpha \in \mathbb{F}_n^+} W_\alpha' [W_1' - T_1' \ \cdots \ W_n' - T_n'] h_\alpha.
\end{aligned}$$

Let $\mathcal{T} := [T_1 \ \cdots \ T_n]$, $T_i \in B(\mathcal{H})$, be an isometry and let $A \in B(\mathcal{H}, \mathcal{H}')$ be an operator such that $\|A\| \leq t$ and $AT_i = T_i' A$ for any $i = 1, \ldots, n$. If B is an intertwining lifting of A with respect to $\mathcal{V}'$ such that $\|B\| \leq t$, then ΦB is an intertwining lifting of A with respect to $\mathcal{W}'$. Since Φ is unique, we call the operator $\tilde{B}_c := \Phi B_c$ the central intertwining lifting of A with respect to $\mathcal{W}'$ and tolerance t. Using these remarks and Theorem 2.4, we can easily obtain the following form for $\tilde{B}_c$.

PROPOSITION 2.6. *Let $\mathcal{T} := [T_1 \ \cdots \ T_n]$, $T_i \in B(\mathcal{H})$, be an isometry and $\mathcal{T}' := [T_1' \ \cdots \ T_n']$, $T_i' \in B(\mathcal{H}')$, be a row contraction. Let $\mathcal{W}' := [W_1' \ \cdots \ W_n']$, $W_i' \in B(\mathcal{G}')$, be an isometric lifting of of $\mathcal{T}'$, and $A \in B(\mathcal{H}, \mathcal{H}')$ be such that $\|A\| \leq t$ and*

$$AT_i = T_i' A, \quad i = 1, \ldots, n.$$

Then the central intertwining lifting $\tilde{B}_c$ of A with respect to $\mathcal{W}'$ and tolerance $t > 0$ is given by

$$\tilde{B}_c = A + \sum_{\alpha \in \mathbb{F}_n^+} W_\alpha' (P' U P_{\mathcal{N}}) E_{\tilde{\sigma}}' D_{A,t},$$

where $E_{g_0} := I_{\mathcal{D}_{A,t}}$, $E_i' := P_i' U P_{\mathcal{N}}$, $i = 1, \ldots, n$, P'(resp. P_i') is the orthogonal projection of $\mathcal{G}' \oplus (\oplus_{j=1}^n \mathcal{D}_{A,t})$ onto $\mathcal{G}'$(resp. the i-th component of $\oplus_{j=1}^n \mathcal{D}_{A,t}$), and *the operator* $U : \mathcal{N} \to \mathcal{G}' \oplus (\oplus_{j=1}^n \mathcal{D}_{A,t})$ *is the isometry defined by*

$$U\left(\sum_{i=1}^n D_{A,t}T_i h_i\right) := \left(\sum_{i=1}^n (W_i' - T_i')Ah_i\right) \oplus (\oplus_{i=1}^n D_{A,t}h_i).$$

At the end of this section, we show that, under certain conditions, there is only one intertwining lifting B of A such that $\|B\| = \|A\|$. We say that an operator $A \in B(\mathcal{H}, \mathcal{H}')$ attains its norm if there is a vector $h \in \mathcal{H}$ of norm one such that $\|Ah\| = \|A\|$.

PROPOSITION 2.7. *Let* $T' := [T_1' \cdots T_n']$, $T_i' \in B(\mathcal{H}')$, *be a row contraction, and let* $A : F^2(H_n) \to \mathcal{H}'$ *be a contraction which attains its norm* $\|A\| = 1$. *If*

$$AS_i = T_i'A, \quad i = 1, \ldots, n,$$

then

$$\mathcal{M} := \mathcal{D}_A \ominus \bigvee_{i=1}^n D_A S_i F^2(H_n) = \{0\}. \tag{2.30}$$

In particular, if $\mathcal{V}' := [V_1' \cdots V_n']$ *is the minimal isometric dilation of* T', *then there is only one intertwining lifting* B *of* A *with respect to* $\mathcal{V}'$ *such that* $\|B\| = \|A\|$.

PROOF. First, note that A attains its norm at a vector $f \in F^2(H_n)$ if and only if $f \in \ker D_A$. We need to prove that A attains its norm at $f \in F^2(H_n)$ such that $f(0) \neq 0$. To this end, assume that A attains its norm at a vector $g \in F^2(H_n)$ such that $g(0) = 0$. Let $g := \sum_{\alpha \in \mathbb{F}_n^+} a_\alpha e_\alpha$ and let k be the least positive integer such that there is $\alpha_0 \in \mathbb{F}_n^+$ with $|\alpha_0| = k$ and $a_{\alpha_0} \neq 0$. Then g can be written as $g = \sum_{|\alpha|=k} S_\alpha \varphi_\alpha$, where $\varphi_\alpha \in F^2(H_n)$. Since $\|g\| = 1$ and $\|Ag\| = \|A\|$, we have

$$\begin{aligned} \|A\|^2 \sum_{|\alpha|=k} \|\varphi_\alpha\|^2 &= \|A\|^2 \|g\|^2 = \|Ag\|^2 \\ &= \left\| A\left(\sum_{|\alpha|=k} S_\alpha \varphi_\alpha\right)\right\|^2 = \left\|\sum_{|\alpha|=k} T_\alpha' A \varphi_\alpha\right\|^2 \\ &\leq \sum_{|\alpha|=k} \|A\varphi_\alpha\|^2 \leq \|A\|^2 \sum_{|\alpha|=k} \|\varphi_\alpha\|^2. \end{aligned}$$

Therefore, we must have

$$\|A\varphi_\alpha\| = \|A\|\|\varphi_\alpha\|, \quad \text{if } |\alpha| = k.$$

Setting $f := \frac{\varphi_{\alpha_0}}{\|\varphi_{\alpha_0}\|}$, we have $\|Af\| = \|A\|$, $\|f\| = 1$, and $f(0) \neq 0$. If we assume that there is $\psi \in \mathcal{M}$, $\psi \neq 0$, then $\langle \psi D_A S_i h\rangle = 0$ for any $h \in F^2(H_n)$ and $i = 1, \ldots, n$. This implies $D_A \psi = a_0 \in \mathbb{C}$, $a_0 \neq 0$. Since $f \in \ker D_A$, we have

$$0 = \langle f, D_A \psi\rangle = \langle f, a_0\rangle.$$

On the other hand, $f(0) \neq 0$ and $a_0 \neq 0$ imply $\langle f, a_0\rangle \neq 0$, which is a contradiction. Therefore, $\mathcal{M} = \{0\}$. The uniqueness of the intertwining lifting follows from Theorem 2.1 and Theorem 2.2, noticing that $C_k = 0$ for any $k = 1, 2, \ldots$. The proof is complete. □

It is easy to see that one can obtain a version of Proposition 2.7 if $\|A\| = t > 0$. In this case, we have to replace D_A by $D_{A,t}$.

2.3. A maximum principle for the noncommutative commutant lifting theorem

In this section, we prove a maximum principle for the noncommutative commutant lifting theorem, which also provides a new characterization for the central intertwining lifting (see Theorem 2.8 and Theorem 2.9). This is a key result used in Section 2.6 to find the maximal entropy solution for the noncommutative commutant lifting theorem.

Let $T_1, \ldots, T_n \in B(\mathcal{H})$ be isometries with orthogonal ranges and let $\mathcal{L} := \bigcap_{i=1}^n \ker T_i^*$ be the wandering subspace in the Wold type decomposition [**38**]. If $X \in B(\mathcal{H}, \mathcal{H}')$ satisfies $\|X\| \le t$ for some $t > 0$, we define the positive operator $\Delta(X) \in B(\mathcal{L})$ by setting

$$\langle \Delta(X)\ell, \ell\rangle := \inf\{\|D_{X,t}(\ell - T_1h_1 - \cdots - T_nh_n)\|^2 : \ h_1, \ldots, h_n \in \mathcal{H}\} \tag{2.31}$$

for any $\ell \in \mathcal{L}$. Following the classical case (see [**23**]), we call the operator $\Delta(X)$ the Schur complement of $D_{X,t}^2$ with respect to $T_1, \ldots, T_n$. Note that if $\|X\| < t$, then $\Delta(X)$ is indeed the Schur complement of $D_{X,t}^2$ with respect to the orthogonal decomposition

$$\mathcal{H} = \mathcal{L} \oplus [T_1\mathcal{H} \oplus \cdots \oplus T_n\mathcal{H}].$$

It is easy to see that

$$\langle \Delta(X)\ell, \ell\rangle := \inf\{\langle D_{X,t}^2 h, h\rangle : \ h \in \mathcal{H} \text{ and } P_{\mathcal{L}}h = \ell\}.$$

On the other hand, one can prove that

$$\Delta(X) = P_{\mathcal{L}} D_{X,t} P_{\mathcal{M}_X} D_{X,t}|\mathcal{L}, \tag{2.32}$$

where

$$\mathcal{M}_X := \mathcal{D}_{X,t} \ominus \bigvee_{i=1}^n D_{X,t}T_i\mathcal{H}$$

and $P_{\mathcal{L}}$, $P_{\mathcal{M}_X}$ are the orthogonal projections onto $\mathcal{L}$ and $\mathcal{M}_X$, respectively. Indeed, we have

$$\begin{aligned}\langle \Delta(X)\ell, \ell\rangle &= \inf\{\|D_{X,t}(\ell - T_1h_1 - \cdots - T_nh_n)\|^2 : \ h_1, \ldots, h_n \in \mathcal{H}\}\\ &= \inf\{\|D_{X,t}\ell - k\|^2 : \ k \in \bigvee_{i=1}^n D_{X,t}T_i\mathcal{H}\}\\ &= \|P_{\mathcal{M}_X} D_{X,t}\ell\|^2 = \langle (P_{\mathcal{L}} D_{X,t} P_{\mathcal{M}_X} D_{X,t}|\mathcal{L})\ell, \ell\rangle\end{aligned}$$

for any $\ell \in \mathcal{L}$. Hence, we obtain relation (2.32).

We can prove now a maximum principle for the noncommutative commutant lifting theorem. This result also provides a new characterization for the central intertwining lifting.

THEOREM 2.8. *Let $A \in B(\mathcal{H}, \mathcal{H}')$ be an operator satisfying $\|A\| \le t$ and $AT_i = T_i'A$ for any $i = 1, \ldots, n$, where $\mathcal{T} := [T_1 \ \cdots \ T_n]$, $T_i \in B(\mathcal{H})$, is an isometry and $\mathcal{T}' := [T_1' \ \cdots \ T_n']$, $T_i' \in B(\mathcal{H}')$, is a row contraction. Let $\mathcal{V}'$ be the minimal* *isometric dilation of T' and let B_c be the central intertwining lifting of A with respect to $\mathcal{V}'$. Then*

$$\Delta(A) = \Delta(B_c) \geq \Delta(B) \tag{2.33}$$

for any intertwining lifting B of A with tolerance t. Moreover, $\Delta(B_c) = \Delta(B)$ if and only if $B_c = B$.

PROOF. According to Theorem 2.2, for each $k = 0, 1, \ldots$, there exists a unitary operator $\Omega_k : \mathcal{D}_{A_{k+1},t} \to \left(\bigoplus_{j=1}^{n} \mathcal{D}_{A_k,t}\right) \oplus \mathcal{D}_{C_k}$ such that

$$P_{i,k} W_k P_{\mathcal{N}_k} \left(\sum_{j=1}^{n} D_{A_k,t} T_j h_j \right) = D_{A_k,t} h_i, \quad i = 1, \ldots, n.$$

Hence, and using relation (2.16), we have

$$\Omega_k D_{A_{k+1},t} \left(\sum_{j=1}^{n} T_j h_j \right) = D_{A_k,t} h_1 \oplus D_{A_k,t} h_2 \oplus \cdots \oplus D_{A_k,t} h_n.$$

Therefore, we have $\Omega_k \mathcal{N}_{k+1} = \oplus_{j=1}^{n} \mathcal{D}_{A_k,t} \oplus \{0\}$ and $\Omega_k \mathcal{M}_{k+1} = \{0\} \oplus \mathcal{D}_{C_k}$. Using again Theorem 2.2, we infer that

$$\|\Omega_k P_{\mathcal{M}_{k+1}} D_{A_{k+1},t} x\| = \|D_{C_k} P_{\mathcal{M}_k} D_{A_k,t} x\|$$

for any $x \in \mathcal{H}$. This implies

$$\|P_{\mathcal{M}_k} D_{A_k,t} x\| \geq \|D_{C_k} P_{\mathcal{M}_k} D_{A_k,t} x\| = \|P_{\mathcal{M}_{k+1}} D_{A_{k+1},t} x\|, \quad x \in \mathcal{H}. \tag{2.34}$$

Moreover, the equality holds for any $x \in \mathcal{H}$ if and only if $C_k = 0$ for any $k = 0, 1, \ldots$. Hence, we deduce that

$$\|P_{\mathcal{M}} D_{A,t} x\| \geq \|P_{\mathcal{M}_1} D_{A_1,t} x\| \geq \cdots \geq \|P_{\mathcal{M}_k} D_{A_k,t} x\|. \tag{2.35}$$

Now, let $x \in \mathcal{H}$ and note that

$$\|D_{A_k,t} x\|^2 = t^2 \|x\|^2 - \|A_k x\|^2 = \|D_{B,t} x\|^2 + \|(I - P_k) B x\|^2. \tag{2.36}$$

Hence, $\|D_{A_k,t} x\|^2 \geq \|D_{B,t} x\|^2$, which implies

$$\|P_{\mathcal{M}_B} D_{B,t} h\| \leq \left\| D_{B,t} h - D_{B,t} \left(\sum_{j=1}^{n} T_j h_j \right) \right\| \leq \left\| D_{A_k,t} h - D_{A_k,t} \left(\sum_{j=1}^{n} T_j h_j \right) \right\|$$

for any $h_1, \ldots, h_n \in \mathcal{H}$. Taking the infimum over $h_1, \ldots, h_n \in \mathcal{H}$, we obtain

$$\|P_{\mathcal{M}_B} D_{B,t} h\| \leq \|P_{\mathcal{M}_k} D_{A_k,t} h\|, \quad h \in \mathcal{H}. \tag{2.37}$$

Combining (2.35) with (2.37), we infer that

$$\|P_{\mathcal{M}_B} D_{B,t} h\| \leq \|P_{\mathcal{M}} D_{A,t} h\|, \quad h \in \mathcal{H},$$

and $\|P_{\mathcal{M}_B} D_{B,t} h\| = \|P_{\mathcal{M}} D_{A,t} h\|$ for any $h \in \mathcal{H}$ if and only if $C_k = 0$ for any $k = 0, 1, \ldots$, which means that $B = B_c$, the central intertwining lifting of A with tolerance t. Taking into account relation (2.32) and the fact that

$$P_{\mathcal{M}_k} D_{A_k,t} y = P_{\mathcal{M}_B} D_{B,t} y = 0$$

for any $y \in \bigvee_{j=1}^{n} T_j \mathcal{H}$, we complete the proof of the theorem. □

Let us remark that if $\{A_k\}_{k=0}^{\infty}$ is the sequence of k-step intertwining liftings of A corresponding to B, then

$$\Delta(B) = \text{SOT} - \lim_{k\to\infty} \Delta(A_k). \tag{2.38}$$

Indeed, according to the proof of Theorem 2.8, the sequence $\{D_{A_k,t}P_{\mathcal{M}_k}D_{A_k,t}\}_{k=0}^{\infty}$ of positive operators is a decreasing. Therefore,

$$Q := \text{SOT} - \lim_{k\to\infty} D_{A_k,t}P_{\mathcal{M}_k}D_{A_k,t}$$

exists. Using relation (2.37), we infer that

$$D_{B,t}P_{\mathcal{M}_B}D_{B,t} \le Q \le D_{A_k,t}P_{\mathcal{M}_k}D_{A_k,t} \tag{2.39}$$

for any $k = 0, 1, \ldots$. On the other hand, using relation (2.36), we obtain

$$\begin{aligned}\langle Qx, x\rangle \le \|P_{\mathcal{M}_k}D_{A_k,t}x\|^2 &\le \left\|D_{A_k,t}x - \sum_{j=1}^n D_{A_k,t}T_jh_j\right\|^2 \\ &= \left\|D_{B,t}\left(x - \sum_{j=1}^n T_jh_j\right)\right\|^2 + \left\|(I - P_k)B\left(x - \sum_{j=1}^n T_jh_j\right)\right\|^2\end{aligned}$$

for any $k = 0, 1, \ldots$, and $h_1, \ldots, h_n \in \mathcal{H}$. Since $P_k \to I$ strongly, as $k \to \infty$, we obtain

$$\langle Qx, x\rangle \le \left\|D_{B,t}\left(x - \sum_{j=1}^n T_jh_j\right)\right\|^2.$$

Taking the infimum over $h_1, \ldots, h_n \in \mathcal{H}$ we get $Q \le D_{B,t}P_{\mathcal{M}_B}D_{B,t}$, which together with relation (2.39) imply $Q = D_{B,t}P_{\mathcal{M}_B}D_{B,t}$. Now, it is clear that relation (2.38) holds.

Does Theorem 2.8 remain true if $\mathcal{V}'$ is an arbitrary isometric lifting of $\mathcal{T}'$? The answer is given in what follows.

THEOREM 2.9. *Let* $\mathcal{T} := [T_1 \ \cdots \ T_n]$, $T_i \in B(\mathcal{H})$, *be a row isometry and* $\mathcal{T}' := [T_1' \ \cdots \ T_n']$, $T_i' \in B(\mathcal{H}')$, *be a row contraction. Let* $\mathcal{W}' := [W_1' \ \cdots \ W_n']$, $V_i' \in B(\mathcal{G}')$, *be an isometric lifting of* T', *and let* $A \in B(\mathcal{H}, \mathcal{H}')$ *be such that* $\|A\| \le t$ *and*

$$AT_i = T_i'A, \quad i = 1, \ldots, n.$$

If B *is an intertwining lifting of* A *with respect to* $\mathcal{W}'$ *and tolerance* $t > 0$, *then*

$$\Delta(B) \le \Delta(\tilde{B}_c) = \Delta(A), \tag{2.40}$$

where $\tilde{B}_c$ *is the central intertwining lifting of* A *with tolerance* $t > 0$. *Moreover,* $\Delta(B) = \Delta(A)$ *if and only if* $\Gamma|\mathcal{M} = 0$, *where* $\Gamma : \mathcal{D}_{A,t} \to \mathcal{G}' \ominus \mathcal{H}'$ *is the unique contraction in the matrix decomposition* $B = \begin{bmatrix} A \\ \Gamma D_{A,t}\end{bmatrix}$.

PROOF. Since B is an intertwining lifting of A with tolerance $t > 0$, we have the matrix representation $B = \begin{bmatrix} A \\ \Gamma D_{A,t}\end{bmatrix}$, where $\Gamma : \mathcal{D}_{A,t} \to \mathcal{G}' \ominus \mathcal{H}'$ is a contraction.

Since $\|D_{A,t}h\| = \|D_\Gamma D_{A,t}h\|$, $h \in \mathcal{H}$, there is a unitary operator $U_\Gamma : \mathcal{D}_{B,t} \to \mathcal{D}_\Gamma$ such that

$$U_\Gamma D_{B,t} = D_\Gamma D_{A,t}. \tag{2.41}$$

Note that, for any $\ell \in \mathcal{L} := \bigcap_{i=1}^n \ker T_i^*$, we have

$$\begin{aligned}\langle \Delta(B)\ell, \ell\rangle &= \inf\left\{\|D_{B,t}(\ell - T_1h_1 - \cdots - T_nh_n)\|^2 :\ h_i \in \mathcal{H}\right\}\\ &= \inf\left\{\|D_\Gamma D_{A,t}(\ell - T_1h_1 - \cdots - T_nh_n)\|^2 :\ h_i \in \mathcal{H}\right\}\\ &\le \inf\left\{\|D_{A,t}(\ell - T_1h_1 - \cdots - T_nh_n)\|^2 :\ h_i \in \mathcal{H}\right\}\\ &= \langle \Delta(A)\ell, \ell\rangle .\end{aligned}$$

Therefore, we obtain

$$\Delta(B) \le \Delta(A). \tag{2.42}$$

On the other hand, since $D_{\tilde{B}_c,t} = D_{B_c,t}$, Theorem 2.8 implies

$$\Delta(\tilde{B}_c) = \Delta(B_c) = \Delta(A).$$

Let $B = \begin{bmatrix} A \\ \Gamma D_{A,t}\end{bmatrix}$ be an intertwining lifting of A with tolerance $t > 0$, and assume that $\Gamma|\mathcal{M} = 0$. Then $D_\Gamma^2|\mathcal{M} = I_\mathcal{M}$, which implies $D_\Gamma|\mathcal{M} = I_\mathcal{M}$ and $\mathcal{M}$ is a reducing subspace for D_Γ. Hence, and taking into account that

$$\mathcal{N} = \bigvee_{i=1}^n D_{A,t}T_i\mathcal{H}, \quad D_\Gamma\mathcal{N} \subset \mathcal{N}, \quad \text{and } D_\Gamma|\mathcal{M} = I_\mathcal{M},$$

we obtain

$$\begin{aligned}\langle \Delta(B)\ell, \ell\rangle &= \inf\left\{\|D_\Gamma D_{A,t}(\ell - T_1h_1 - \cdots - T_nh_n)\|^2 :\ h_i \in \mathcal{H}\right\}\\ &= \inf\left\{\|D_\Gamma D_{A,t}\ell - D_\Gamma k\| :\ k \in \mathcal{N}\right\}\\ &\ge \inf\left\{\|D_\Gamma D_{A,t}\ell - x\| :\ x \in \mathcal{N}\right\}\\ &= \|P_\mathcal{M} D_\Gamma D_{A,t}\ell\|^2 = \|D_\Gamma P_\mathcal{M} D_{A,t}\ell\|^2\\ &= \|P_\mathcal{M} D_{A,t}\ell\|^2 = \langle \Delta(A)\ell, \ell\rangle .\end{aligned}$$

Therefore, $\Delta(B) \ge \Delta(A)$, which together with relation (2.42) imply $\Delta(B) = \Delta(A)$.

Now, let $B = \begin{bmatrix} A \\ \Gamma D_{A,t}\end{bmatrix}$ be an intertwining lifting of A with tolerance $t > 0$, and assume that $\Delta(B) = \Delta(A)$. Since $\mathcal{N} = \bigvee_{i=1}^n D_{A,t}T_i\mathcal{H}$ and $\mathcal{D}_{A,t} = \mathcal{N} \oplus \mathcal{M}$, we have

$$\begin{aligned}\langle \Delta(B)\ell, \ell\rangle &= \inf\left\{\|D_\Gamma D_{A,t}(\ell - T_1h_1 - \cdots - T_nh_n)\|^2 :\ h_i \in \mathcal{H}\right\}\\ &= \|D_\Gamma P_\mathcal{M} D_{A,t}\ell\|^2\\ &\le \|P_\mathcal{M} D_{A,t}\ell\|^2 = \langle \Delta(A)\ell, \ell\rangle\end{aligned}$$

for any $\ell \in \mathcal{L}$. Since $\Delta(B) = \Delta(A)$, we infer that

$$\|D_\Gamma P_\mathcal{M} D_{A,t}\ell\| = \|P_\mathcal{M} D_{A,t}\ell\|, \quad \ell \in \mathcal{L}.$$

Hence, we get $\Gamma|\mathcal{M} = 0$. The proof is complete. □

2.4. A permanence principle for the central intertwining lifting

In this section, we present a permanence principle for the central intertwining lifting. This generalizes the permanence principle for the Carathéodory interpolation problem in [**19**] (case $n = 1$) to our multivariable setting. Applications of this principle will be considered in the next sections.

Let $[T_1' \cdots T_n']$, $T_i' \in B(\mathcal{H}')$, be a row contraction and let $[V_1' \cdots V_n']$, $V_i' \in B(\mathcal{K}')$, be its minimal isometric dilation. Let $[T_1 \cdots T_n]$, $T_i \in B(\mathcal{H})$, be a row isometry and let $\mathcal{M}' \subseteq \mathcal{K}'$ be an invariant subspace under each $V_i'^*$, $i = 1, \ldots, n$, such that $\mathcal{H}' \subseteq \mathcal{M}'$. Let $A \in B(\mathcal{H}, \mathcal{H}')$ be such that $\|A\| \leq t$ and $AT_i = T_i'A$ for any $i = 1, \ldots, n$. An operator $Y : \mathcal{H} \to \mathcal{M}'$ is called partial intertwining lifting of A if

$$P_{\mathcal{M}'}Y = A \quad \text{and} \quad YT_i = (P_{\mathcal{M}'}V_i'|\mathcal{M}')Y, \quad i = 1, \ldots, n. \tag{2.43}$$

For example, a k-step intertwining lifting A_k is a partial intertwining lifting of A.

THEOREM 2.10. *Let* $[T_1 \cdots T_n]$, $T_i \in B(\mathcal{H})$, *be a row isometry and let* $[T_1' \cdots T_n']$, $T_i' \in B(\mathcal{H}')$, *be a row contraction with minimal isometric dilation* $\mathcal{V}'$. *If* $A : \mathcal{H} \to \mathcal{H}'$ *is such that* $\|A\| \leq t$ *and*

$$AT_i = T_i'A, \quad i = 1, \ldots, n,$$

and $Y : \mathcal{H} \to \mathcal{M}'$ *is a contractive partial intertwining lifting of* A, *then*

$$\Delta(A) \geq \Delta(Y). \tag{2.44}$$

The equality holds if and only if $Y = P_{\mathcal{M}'}B_c$, *where* B_c *is the central intertwining lifting of* A *with respect to* $\mathcal{V}'$ *and tolerance* t. *Moreover, the central intertwining lifting of* $P_{\mathcal{M}'}B_c$ *is precisely the central intertwining lifting of* A.

PROOF. First note that $[V_1' \cdots V_n']$ is also the minimal isometric dilation of the row contraction $[P_{\mathcal{M}'}V_1'|\mathcal{M}' \cdots P_{\mathcal{M}'}V_n'|\mathcal{M}']$. This follows from the fact that $\mathcal{H}' \subset \mathcal{M}'$, $V_i'^*\mathcal{M}' \subset \mathcal{M}'$ for any $i = 1, \ldots, n$, and the minimality condition $\mathcal{K}' = \bigvee_{\alpha \in \mathbb{F}_n^+} V_\alpha'\mathcal{H}'$. On the other hand, any intertwining lifting $\hat{Y}$ of Y is also an intertwining lifting of A. Indeed, we have $V_i'\hat{Y} = \hat{Y}T_i$, $i = 1, \ldots, n$, and

$$P_{\mathcal{H}'}\hat{Y} = P_{\mathcal{H}'}P_{\mathcal{M}'}\hat{Y} = P_{\mathcal{H}'}Y = A.$$

Now, let Y_c be the central intertwining of Y. Applying Theorem 2.8, we get $\Delta(Y) = \Delta(Y_c)$. Since Y_c is an intertwining lifting of A, we can use again Theorem 2.8 to obtain

$$\Delta(A) \geq \Delta(Y_c).$$

Therefore, we infer relation (2.44). According to Theorem 2.8, the equality $\Delta(A) = \Delta(Y_c)$ holds if and only if $Y_c = B_c$, where B_c is the central intertwining of A. Hence, $Y = P_{\mathcal{M}'}Y_c = P_{\mathcal{M}'}B_c$.

To prove the second part of the theorem, let $Y = P_{\mathcal{M}'}B_c$. Since $\mathcal{M}'$ is an invariant subspace under each V_i^*, $i = 1, \ldots, n$, we have

$$\begin{aligned}(P_{\mathcal{M}'}V_i'|\mathcal{M}')Y &= P_{\mathcal{M}'}V_1'P_{\mathcal{M}'}B_c = P_{\mathcal{M}'}V_1'B_c \\ &= P_{\mathcal{M}'}B_cT_i = YT_i,\end{aligned}$$

for any $i = 1, \ldots, n$. Hence, Y is a partial intertwining lifting of A. Applying the first part of the theorem to the partial intertwining lifting Y, we have

$$\Delta(A) \geq \Delta(Y) = \Delta(Y_c), \tag{2.45}$$

where Y_c is the central intertwining lifting of Y. Since B_c is an intertwining lifting of Y, Theorem 2.8 implies

$$\Delta(Y_c) \geq \Delta(B_c) = \Delta(A). \tag{2.46}$$

Combining relation (2.45) with (2.46), we get $\Delta(Y_c) = \Delta(B_c)$. Since Y_c and B_c are intertwining liftings of A, the uniqueness part of Theorem 2.8 implies $Y_c = B_c$. The proof is complete. □

2.5. Quasi outer spectral factorizations

In this section, we obtain explicit formulas for the quasi outer spectral factor of the defect operator $t^2 I - B_c^* B_c$ of the central intertwining lifting B_c (see Theorem 2.14). This leads, in the next section, to concrete formulas for the entropy of B_c as well as to a maximum principle and a characterization of the central intertwining lifting $\tilde{B}_c$ with respect to non-minimal isometric liftings.

Let $\mathcal{T} := [T_1 \cdots T_n]$, $T_i \in B(\mathcal{H})$, be a row isometry, $\mathcal{T}' := [T_1' \cdots T_n']$, $T_i' \in B(\mathcal{H}')$, be a row contraction, and let $A : \mathcal{H} \to \mathcal{H}'$ be such that $\|A\| \leq t$ and

$$AT_i = T_i' A, \quad i = 1, \ldots, n.$$

Define the operator $X_A \in B(\oplus_{i=1}^n \mathcal{H}, \mathcal{H})$ by

$$X^A := D_{A,t}^2 [T_1 \cdots T_n][T_i^* D_{A,t}^2 T_j]_{n\times n}, \tag{2.47}$$

and let $X^A = [X_1^A \cdots X_n^A]$ be its matrix representation with $X_i^A \in B(\mathcal{H})$, $i = 1, \ldots, n$. As is [**55**], one can prove that if $\|A\| < t$, then the central intertwining lifting of A satisfies the equation

$$B_c h = Ah \oplus \sum_{\sigma \in \mathbb{F}_n^+} e_\sigma \otimes D_{\mathcal{T}'}(\oplus_{i=1}^n A)(X_\sigma^A)^* h \tag{2.48}$$

for any $h \in \mathcal{H}$.

LEMMA 2.11. *If $\|A\| < t$, then the following statements hold:*

(i) $\mathcal{M} := \mathcal{D}_{A,t} \ominus \mathcal{N} = D_{A,t}^{-1}\mathcal{L}$, *where $\mathcal{L} := \bigcap_{i=1}^n \ker T_i^*$ and $\mathcal{N}$ is defined by* (2.13) (*when $k = 0$*);

(ii) $B_c D_{A,t}^{-2}\ell = A D_{A,t}^{-2}\ell$ *for any $\ell \in \mathcal{L}$;*

(iii) $E_i = D_{A,t}(X_i^A)^* D_{A,t}^{-1}$, *where $E_i := P_i W P_{\mathcal{N}}$ for any $i = 1, \ldots, n$, and the operator W is defined by* (2.14) (*when $k = 0$*).

PROOF. To prove (i), note that $h \in \mathcal{M}$ if and only if $h \perp \bigvee_{i=1}^n D_{A,t} T_i \mathcal{H}$, which is equivalent to $D_{A,t} h \in \bigcap_{i=1}^n \ker T_i^*$. Hence, $h \in D_{A,t}^{-1}\mathcal{L}$. The statement (ii) follows from relation (2.48) and the fact that, for any $\ell \in \bigcap_{i=1}^n \ker T_i^*$,

$$(X^A)^* D_{A,t}^{-2}\ell = [T_j^* D_{A,t}^2 T_i]_{n\times n} \begin{bmatrix} T_1^* \\ \vdots \\ T_n^* \end{bmatrix} D_{A,t}^2 (D_{A,t}^{-2}\ell) = 0.$$

The proof of the similarity relation (iii) is exactly the same as that from [**55**] for the case $t = 1$. We shall omit it. □

We mention that part (i) can be used together with Theorem 2.4 to give another proof for part (ii) of Lemma 2.11. We recall that $\mathcal{T} := [T_1 \cdots T_n]$, $T_i \in B(\mathcal{H})$, is a C_0-row contraction if

$$\lim_{k\to\infty} \sum_{|\alpha|=k} \|T_\alpha^* h\|^2 = 0$$

for any $h \in \mathcal{H}$. If $\mathcal{T} = [S_1 \otimes I_{\mathcal{M}} \cdots S_n \otimes I_{\mathcal{M}}]$ for some Hilbert space $\mathcal{M}$, then $\mathcal{T}$ is called orthogonal shift of multiplicity $\dim \mathcal{M}$.

LEMMA 2.12. *Let $\mathcal{T} := [T_1 \cdots T_n]$, $T_i \in B(\mathcal{H})$, be a row isometry, $\mathcal{T}' := [T_1' \cdots T_n']$, $T_i' \in B(\mathcal{H}')$, be a row contraction, and let $A : \mathcal{H} \to \mathcal{H}'$ be such that $\|A\| \le t$ and*

$$AT_i = T_i' A, \quad i = 1, \ldots, n.$$

Then there exists a unique positive operator $Q \in B(\mathcal{D}_{A,t})$ such that

$$\|QD_{A,t}h\|^2 = \lim_{k\to\infty} \sum_{|\sigma|=k} \|E_\sigma D_{A,t}h\|^2, \quad h \in \mathcal{H}. \tag{2.49}$$

Moreover, if $\mathcal{T}$ is an orthogonal shift and $\|A\| < t$, then $Q = 0$ and $[X_1^A \cdots X_n^A]$ is a C_0-row contraction.

PROOF. Note that if $E_i := P_i W P_{\mathcal{N}}$, $i = 1, \ldots, n$, then $[E_1^* \cdots E_n^*]$ is a row contraction from $\oplus_{i=1}^n \mathcal{D}_{A,t}$ to $\mathcal{D}_{A,t}$. Since $\{\sum_{|\alpha|=k} E_\alpha^* E_\alpha\}_{k=1}^\infty$ is a decreasing sequence of contractions, we can define the operator

$$Q := \left(\text{SOT} - \lim_{k\to\infty} \sum_{|\alpha|=k} E_\alpha^* E_\alpha \right)^{1/2}.$$

Hence, relation (2.49) holds. Now, assume that $\mathcal{T}$ is an orthogonal shift and $\|A\| < 1$. Since, for any $i = 1, \ldots, n$,

$$P_j W P_{\mathcal{N}} D_{A,t} T_i h = \delta_{ij} D_{A,t} h, \quad h \in \mathcal{H}, \tag{2.50}$$

we deduce

$$E_{\tilde{\omega}} D_{A,t} T_\sigma h = \delta_{\sigma,\omega} D_{A,t} h, \tag{2.51}$$

for any $\sigma, \omega \in \mathbb{F}_N^+$. According to Lemma 3.3. from [**55**], the span $\bigvee_{\sigma\in\mathbb{F}_n^+} D_{A,t} T_\sigma D_{A,t}^{-2} \mathcal{L}$ is equal to $\mathcal{H}$, where $\mathcal{L} := \bigcap_{i=1}^n \ker T_i^*$. Taking into account that $\mathcal{D}_{A,t} = \mathcal{N} \oplus \mathcal{M}$ and $\mathcal{M} = D_{A,t}^{-1} \mathcal{L}$, we have

$$\sum_{j=1}^n \sum_{|\alpha|=k} \|E_j E_\alpha D_{A,t} T_\sigma D_{A,t}^{-2} \ell\|^2 = \sum_{j=1}^n \|E_j D_{A,t}^{-1} \ell\|^2 = 0.$$

This shows that $[E_1^* \cdots E_n^*]$ is a C_0-row contraction. Using Lemma 2.11 part (iii), we conclude that X^A is a a C_0-row contraction. □

LEMMA 2.13. *Let $\mathcal{Y}_Q := \overline{\text{range}\, Q}$ and define the operators $V_i \in B(\mathcal{Y}_Q)$, $i = 1, \ldots, n$, by setting*

$$V_i Q D_{A,t} h := Q D_{A,t} T_i h, \quad h \in \mathcal{H}. \tag{2.52}$$

Then $V_1, \ldots, V_n$ are isometries with orthogonal ranges and

$$V_1 V_1^* + \cdots + V_n V_n^* = I_{\mathcal{Y}_Q}.$$

PROOF. Using Lemma 2.12 and relation (2.51), we have

$$\begin{aligned}\|QD_{A,t}T_i h\|^2 &= \lim_{k\to\infty} \sum_{j=1}^n \sum_{|\alpha|=k-1} \|E_\alpha E_j D_{A,t} T_i h\|^2 \\ &= \lim_{k\to\infty} \sum_{|\alpha|=k-1} \|E_\alpha D_{A,t} h\|^2 = \|QD_{A,t}h\|^2\end{aligned}$$

for any $h \in \mathcal{H}$ and $i = 1, \ldots, n$. Similar computations show that

$$\langle QD_{A,t}T_i h, QD_{A,t}T_j h\rangle = \text{SOT} - \lim_{k\to\infty} \left\langle \sum_{|\alpha|=k} E_\alpha^* E_\alpha D_{A,t} T_i h, D_{A,t} T_j h \right\rangle = 0$$

for any $i \neq j$. Therefore, $\{V_i\}_{i=1}^n$ are isometries with orthogonal ranges. On the other hand, since $Q \geq 0$, $\mathcal{N} = \bigvee_{i=1}^n D_{A,t} T_i \mathcal{H}$, and the range of Q is included in $\mathcal{N}$, we infer that

$$\bigvee_{i=1}^n QD_{A,t}T_i\mathcal{H} = \overline{Q\mathcal{D}_{A,t}}.$$

Now, it is clear that $\bigvee_{i=1}^n QD_{A,t}T_i\mathcal{H}$ is dense in $\mathcal{Y}_Q$ and

$$V_1 V_1^* + \cdots + V_n V_n^* = I_{\mathcal{Y}_Q}.$$

□

In the next theorem, we prove that the defect operator $t^2 I - B_c^* B_c$ admits a quasi outer spectral factor, i.e., $t^2 I - B_c^* B_c = Z^* Z$ for some operator Z with dense range.

THEOREM 2.14. *Let $\mathcal{T} := [T_1 \ \cdots \ T_n]$, $T_i \in B(\mathcal{H})$, be a row isometry, $\mathcal{T}' := [T_1' \ \cdots \ T_n']$, $T_i' \in B(\mathcal{H}')$, be a row contraction, and let $A : \mathcal{H} \to \mathcal{H}'$ be such that $\|A\| \leq t$ and*

$$AT_i = T_i' A, \quad i = 1, \ldots, n.$$

Let $Z : \mathcal{H} \to \mathcal{Y}_Q \oplus [F^2(H_n) \otimes \mathcal{M}]$ be defined by

$$Zh := QD_{A,t}h \oplus \sum_{\sigma\in\mathbb{F}_n^+} e_\sigma \otimes P_{\mathcal{M}} E_{\tilde{\sigma}} D_{A,t} h,$$

where $\mathcal{M} := \mathcal{D}_{A,t} \ominus \mathcal{N}$. Then the operator Z has dense range and

$$t^2 I - B_c^* B_c = Z^* Z,$$

where B_c is the central intertwining lifting of A with respect to the minimal isometric dilation of $\mathcal{T}'$ and tolerance t. Moreover, for any $i = 1, \ldots, n$, we have

$$ZT_i = \begin{bmatrix} V_i & 0 \\ 0 & S_i \otimes I_{\mathcal{M}} \end{bmatrix} Z, \quad i = 1, \ldots, n, \tag{2.53}$$

where the isometries $V_i \in B(\mathcal{Y}_Q)$ are defined by relation (2.52).

PROOF. As in the proof of Theorem 2.4, for any $h \in \mathcal{H}$, we have

$$\begin{aligned}\|D_{B_c,t}h\|^2 &= \|D_{A,t}h\|^2 - \|P_{\mathcal{N}}D_{A,t}h\|^2 \\ &\quad + \lim_{k\to\infty}\left(\sum_{1\leq|\omega|\leq k}\|E_{\tilde{\omega}}D_{A,t}h\|^2 - \sum_{1\leq|\sigma|\leq k-1}\|P_{\mathcal{N}}E_{\tilde{\sigma}}D_{A,t}h\|^2\right) \\ &= \|P_{\mathcal{M}}D_{A,t}h\|^2 + \lim_{k\to\infty}\sum_{|\alpha|=k}\|E_{\tilde{\alpha}}D_{A,t}h\|^2 + \sum_{|\sigma|\geq 1}\|P_{\mathcal{M}}E_{\tilde{\sigma}}D_{A,t}h\|^2.\end{aligned}$$

Hence, and using Lemma 2.12, we get

$$\begin{aligned}\|D_{B_c,t}h\|^2 &= \|QD_{A,t}h\|^2 + \sum_{\sigma\in\mathbb{F}_n^+}\|P_{\mathcal{M}}E_{\tilde{\sigma}}D_{A,t}h\|^2 \\ &= \|Zh\|^2\end{aligned}$$

for any $h \in \mathcal{H}$. Therefore, we have $t^2I - B_c^*B_c = Z^*Z$.

Now, we prove the equality (2.53). Using Lemma 2.13 and relation (2.51), we obtain

$$\begin{aligned}\begin{bmatrix} V_i & 0 \\ 0 & S_i\otimes I_{\mathcal{M}}\end{bmatrix} Zh &= V_iQD_{A,t}h \oplus \sum_{\sigma\in\mathbb{F}_n^+} e_{g_i\sigma}\otimes P_{\mathcal{M}}E_{\tilde{\sigma}}D_{A,t}h \\ &= QD_{A,t}T_ih \oplus \sum_{\sigma\in\mathbb{F}_n^+} e_{g_i\sigma}\otimes P_{\mathcal{M}}E_{\tilde{\sigma}}E_iD_{A,t}T_ih \\ &= QD_{A,t}T_ih \oplus \sum_{\sigma\in\mathbb{F}_n^+}\sum_{j=1}^{n} e_{g_j\sigma}\otimes P_{\mathcal{M}}E_{\tilde{\sigma}}E_jD_{A,t}T_ih \\ &= QD_{A,t}T_ih \oplus \sum_{\omega\in\mathbb{F}_n^+} e_{\omega}\otimes P_{\mathcal{M}}E_{\tilde{\omega}}D_{A,t}T_ih \\ &= ZT_ih\end{aligned}$$

for any $h \in \mathcal{H}$ and $i = 1, \ldots, n$.

It remains to show that Z has dense range. Note that $\mathcal{Y}_Q \subseteq \mathcal{N}$ and let $f \oplus \varphi$ be in $\mathcal{Y}_Q \oplus [F^2(H_n)\otimes\mathcal{M}]$ such that $f\oplus\varphi \perp Z\mathcal{H}$. Consider the representation

$$\varphi = \sum_{\alpha\in\mathbb{F}_n^+} e_\alpha\otimes h_\alpha, \quad h_\alpha\in\mathcal{M}.$$

Using the definition of Z, we have

$$\begin{aligned}0 = \langle f\oplus\varphi, Zh\rangle &= \langle f, QD_{A,t}h\rangle + \sum_{\alpha\in\mathbb{F}_n^+}\langle h_\alpha, P_{\mathcal{M}}E_{\tilde{\alpha}}D_{A,t}h\rangle \\ &= \langle D_{A,t}Qf, h\rangle + \sum_{\alpha\in\mathbb{F}_n^+}\langle D_{A,t}E_{\tilde{\alpha}}^*h_\alpha, h\rangle,\end{aligned}$$

for any $h \in \mathcal{H}$. Therefore,

$$D_{A,t}\left[Qf + \sum_{\alpha\in\mathbb{F}_n^+} E_{\tilde{\alpha}}^*h_\alpha\right] = 0.$$

Since $D_{A,t}$ is a one-to-one operator on $\mathcal{D}_{A,t} = \mathcal{N} \oplus \mathcal{M}$ and the range of Q is included in $\mathcal{N}$, it follows that

$$h_{g_0} + Qf + \sum_{|\alpha| \geq 1} E_{\tilde{\alpha}}^* h_\alpha = 0.$$

Since $h_{g_0} \in \mathcal{M}$ and the other summands are in $\mathcal{N}$, we get $h_{g_0} = 0$ and

$$Qf + \sum_{|\alpha| \geq 1} E_{\tilde{\alpha}}^* h_\alpha = 0. \tag{2.54}$$

Now, note that, for any $x \in \mathcal{D}_{A,t}$, we have

$$\sum_{i=1}^n \|Q(P_i W P_{\mathcal{N}})x\|^2 = \lim_{k \to \infty} \sum_{|\sigma|=k} \sum_{i=1}^n \|E_\sigma E_i x\|^2 = \|Qx\|^2. \tag{2.55}$$

Define the operators $W_i \in B(\mathcal{Y}_Q)$, $i = 1, \ldots, n$, by setting

$$W_i Q h := Q E_i h, \quad h \in \mathcal{H}. \tag{2.56}$$

According to relation (2.55), we have

$$W_1^* W_1 + \cdots + W_n^* W_n = I_{\mathcal{Y}_Q}. \tag{2.57}$$

Using relations (2.56) and (2.57), we obtain $QW_i^* W_i = E_i^* Q E_i$, for any $i = 1, \ldots, n$, and $Q = \sum_{i=1}^n E_i^* Q W_i$, which together with relation (2.54) imply

$$\sum_{i=1}^n E_i^* Q W_i f + \sum_{|\alpha| \geq 1} E_{\tilde{\alpha}}^* h_\alpha = 0.$$

This equality can be written as

$$\sum_{i=1}^n E_i^* \left(Q W_i f + h_{g_i} + \sum_{|\sigma| \geq 2} E_{\tilde{\sigma}}^* h_{g_i \sigma} \right) = 0. \tag{2.58}$$

Note that the subspace

$$\{E_1 k \oplus E_2 k \oplus \cdots \oplus E_n k : \quad k \in \mathcal{D}_{A,t}\}$$

is dense in $\oplus_{i=1}^n \mathcal{D}_{A,t}$. Indeed, this is due to the equations

$$E_i D_{A,t} T_j h = \delta_{ij} D_{A,t} h, \quad i, j = 1, \ldots, n.$$

Therefore, the operator $[E_1^* \ \cdots \ E_n^*]$ is injective and relation (2.58) implies

$$Q W_i f + h_{g_i} + \sum_{|\sigma| \geq 2} E_{\tilde{\sigma}}^* h_{g_i \sigma} = 0, \quad i = 1, \ldots, n.$$

Since $h_{g_i} \in \mathcal{M}$ and the other summands are in $\mathcal{N}$, we obtain $h_{g_i} = 0$, for any $i = 1, \ldots, n$. Continuing this process, one can prove that $h_\alpha = 0$ for any $\alpha \in \mathbb{F}_n^+$. Therefore, relation (2.54) implies $Qf = 0$. Since f is in the range of Q, we get $f = 0$. The proof is complete. □

COROLLARY 2.15. *Let $\mathcal{T} := [T_1 \ \cdots \ T_n]$, $T_i \in B(\mathcal{H})$, be a row isometry and $\mathcal{L} := \bigcap_{i=1}^n \ker T_i^*$. Let $\mathcal{T}' := [T_1' \ \cdots \ T_n']$, $T_i' \in B(\mathcal{H}')$, be a row contraction, and let $A : \mathcal{H} \to \mathcal{H}'$ be such that $\|A\| < t$ and*

$$A T_i = T_i' A, \quad i = 1, \ldots, n.$$

Let $Z_{\mathcal{L}} : \mathcal{H} \to \mathcal{Y}_Q \oplus [F^2(H_n) \otimes \mathcal{L}]$ *be defined by*

$$Z_{\mathcal{L}} h := Q D_{A,t} h \oplus \sum_{\sigma \in \mathbb{F}_n^+} e_\sigma \otimes N_A P_{\mathcal{L}} (X_\sigma^A)^* h,$$

where $N_A := (P_{\mathcal{L}} D_{A,t}^{-2} | \mathcal{L})^{-1/2}$ *and* X^A *is defined by relation* (2.47). *Then* $Z_{\mathcal{L}}$ *has dense range and*

$$t^2 I - B_c^* B_c = Z_{\mathcal{L}}^* Z_{\mathcal{L}},$$

where B_c *is the central intertwining lifting of* A *with tolerance* t. *Moreover, for any* $i = 1, \ldots, n$, *we have*

$$Z_{\mathcal{L}} T_i = \begin{bmatrix} V_i & 0 \\ 0 & S_i \otimes I_{\mathcal{L}} \end{bmatrix} Z_{\mathcal{L}}, \quad i = 1, \ldots, n, \tag{2.59}$$

where the isometries $V_i \in B(\mathcal{Y}_Q)$ *are defined by relation* (2.52).

Proof. According to Lemma 2.11, we have $\mathcal{M} = \mathcal{D}_{A,t} \ominus \mathcal{N} = D_{A,t}^{-1} \mathcal{L}$, where $\mathcal{L} = \bigcap_{i=1}^n \ker T_i^*$. Note that the operator $X \in B(\mathcal{L}, \mathcal{H})$ defined by $X := D_{A,t}^{-1} | \mathcal{L}$ is injective and has range equal to $\mathcal{M}$. Hence, one can easily see that

$$P_{\mathcal{M}} = X(X^* X)^{-1} X^* = (D_{A,t}^{-1} | \mathcal{L})(P_{\mathcal{L}} D_{A,t}^{-2} | \mathcal{L})^{-1} P_{\mathcal{L}} D_{A,t}^{-1}. \tag{2.60}$$

Using again part (iii) of Lemma 2.11 and relation (2.60), we obtain

$$\begin{aligned} \sum_{\sigma \in \mathbb{F}_n^+} e_\sigma \otimes P_{\mathcal{M}} E_{\tilde{\sigma}} D_{A,t} h &= \sum_{\sigma \in \mathbb{F}_n^+} e_\sigma \otimes P_{\mathcal{M}} D_{A,t} (X_\sigma^A)^* h \\ &= \sum_{\sigma \in \mathbb{F}_n^+} e_\sigma \otimes \Omega N_A P_{\mathcal{L}} (X_\sigma^A)^* h \\ &= (I_{F^2(H_n)} \otimes \Omega) \left(\sum_{\sigma \in \mathbb{F}_n^+} e_\sigma \otimes N_A P_{\mathcal{L}} (X_\sigma^A)^* h \right), \end{aligned}$$

where

$$\Omega := (D_{A,t}^{-1} | \mathcal{L})(P_{\mathcal{L}} D_{A,t}^{-2} | \mathcal{L})^{-1/2} \quad \text{and} \quad N_A := (P_{\mathcal{L}} D_{A,t}^{-2} | \mathcal{L})^{-1/2}.$$

Note that Ω is a unitary operator from $\mathcal{L}$ onto $\mathcal{M}$. Using Theorem 2.14, we deduce

$$Z = \begin{bmatrix} I_{\mathcal{Y}_Q} & 0 \\ 0 & I_{F^2(H_n)} \otimes \Omega \end{bmatrix} Z_{\mathcal{L}},$$

which completes the proof. □

Corollary 2.16. *Let* $[T_1' \cdots T_n']$, $T_i' \in B(\mathcal{H}')$, *be a row contraction, let* $A : F^2(H_n) \otimes \mathcal{E} \to \mathcal{H}'$ *be such that* $\|A\| < t$ *and*

$$A(S_i \otimes I_{\mathcal{E}}) = T_i' A, \quad i = 1, \ldots, n.$$

If B_c *is the central intertwining lifting of* A *with* $\|B_c\| \leq t$, *then the multi-Toeplitz operator* $t^2 I - B_c^* B_c$ *admits an outer spectral factorization*

$$t^2 I - B_c^* B_c = M_\varphi^* M_\varphi,$$

where $M_\varphi \in R_n^\infty \bar{\otimes} B(\mathcal{E})$ *is an outer multi-analytic operator operator with the symbol* φ *given by*

$$\varphi(h) := \sum_{\sigma \in \mathbb{F}_n^+} e_\sigma \otimes \Delta(A)^{1/2} P_{\mathcal{E}} (X_\sigma^A)^* h, \qquad h \in \mathcal{E}, \tag{2.61}$$

where $\Delta(A) = P_{\mathcal{E}}(t^2 I - A^*A)^{-1}|\mathcal{E}$.

PROOF. Using Lemma 2.12 when $T_i = S_i \otimes I_{\mathcal{E}},\ i = 1, \ldots, n$, we get $Q = 0$. In this particular case, Corollary 2.15 implies

$$Z_{\mathcal{E}}(S_i \otimes I_{\mathcal{E}}) = (S_i \otimes I_{\mathcal{E}})Z_{\mathcal{E}},\ i = 1, \ldots, n.$$

Hence, $Z_{\mathcal{E}} = M_\varphi$ for some $M_\varphi \in R_n^\infty \bar{\otimes} B(\mathcal{E})$. Since $Z_{\mathcal{E}}$ has dense range in $F^2(H_n) \otimes \mathcal{E}$, the operator M_φ is outer. Using again Corollary 2.15, one can easily complete the proof. $\square$

We remark that the multi-analytic operator M_φ in Corollary 2.16 satisfies the equation

$$M_\varphi = (I_{F^2(H_n)} \otimes \Delta(A)^{1/2} P_{\mathcal{E}})(I - R_1 \otimes (X_1^A)^* - \cdots - R_n \otimes (X_n^A)^*)^{-1}.$$

Using relation (2.61), we can deduce this formula as in the proof of Remark 3.10.

2.6. Noncommutative commutant lifting theorem and the maximal entropy solution

In this section, we prove that the central intertwining lifting is the maximal entropy solution for the noncommutative commutant lifting theorem when $\mathcal{T} := [S_1 \otimes I_{\mathcal{E}} \cdots S_n \otimes I_{\mathcal{E}}]$ with $\dim \mathcal{E} < \infty$ (see Theorem 2.17). Based on several results of this paper, we are led to concrete formulas for the entropy of B_c (see Theorem 2.18) and, under a certain condition of stability, to a maximum principle and a characterization (in terms of entropy) of the central intertwining lifting $\tilde{B}_c$ with respect to non-minimal isometric liftings (see Theorem 2.20 and Corollary 2.21).

Let $[T_1' \cdots T_n']$, $T_i' \in B(\mathcal{H}')$, be a row contraction and let $[V_1' \cdots V_n']$, $V_i' \in B(\mathcal{K}')$, be its minimal isometric dilation. Let $A : F^2(H_n) \otimes \mathcal{E} \to \mathcal{H}'$ be such that $\|A\| \leq t$ and

$$A(S_i \otimes I_{\mathcal{E}}) = T_i' A, \quad i = 1, \ldots, n,$$

and let $B : F^2(H_n) \otimes \mathcal{E} \to \mathcal{K}'$ be an intertwining lifting of A satisfying $\|B\| \leq t$. Then B is a generalized multiplier and it makes sense to define its prediction entropy as in Section 1.4, i.e., $E(B) := e(D_{B,t}^2)$, where $D_{B,t}^2 = t^2 I - B^*B$.

In what follows, we show that the central intertwining lifting of A with tolerance $t > 0$ is the maximal entropy intertwining lifting of A. More precisely, we can prove the following result.

THEOREM 2.17. *Let* $[T_1' \cdots T_n']$, $T_i' \in B(\mathcal{H}')$, *be a row contraction with minimal isometric dilation* $\mathcal{V}'$, *and let* $A : F^2(H_n) \otimes \mathcal{E} \to \mathcal{H}'$ *be such that* $\|A\| \leq t$ *and*

$$A(S_i \otimes I_{\mathcal{E}}) = T_i' A, \quad i = 1, \ldots, n.$$

If $\dim \mathcal{E} < \infty$ *and* B_c *is the central intertwining lifting of* A *with respect to* $\mathcal{V}'$ *and tolerance* $t > 0$, *then*

$$E(B_c) \geq E(B),$$

for any intertwining lifting B *of* A. *Moreover, if the entropy* $E(B_c) > -\infty$, *then*

$$E(B_c) = E(B)$$

if and only if $B_c = B$.

PROOF. Let $B : F^2(H_n) \otimes \mathcal{E} \to \mathcal{K}'$ be an intertwining lifting of A satisfying $\|B\| \le t$. Using relations (1.45), (1.2), (1.13), and (2.32) in our setting, we deduce

$$\begin{aligned} E(B) = e(D_{B,t}^2) &= \ln\det \Delta_{D_{B,t}^2} \\ &= \ln\det[P_{\mathcal{E}} D_{B,t} P_{\mathcal{M}} D_{B,t}|\mathcal{E}] \\ &= \ln\det \Delta(B), \end{aligned}$$

where

$$\mathcal{M} := \overline{D_{B,t}(F^2(H_n) \otimes \mathcal{E})} \ominus \bigvee_{i=1}^{n} D_{B,t}(S_i \otimes I_{\mathcal{E}})(F^2(H_n) \otimes \mathcal{E}).$$

Similarly, we obtain $E(B_c) = \ln\det \Delta(B_c)$, where B_c is the central intertwining lifting of A. According to Theorem 2.8, we have $\Delta(B_c) \ge \Delta(B)$. Therefore, we get

$$E(B_c) = \ln\det \Delta(B_c) \ge \ln\det \Delta(B) = E(B).$$

Now, assume that $E(B_c) > -\infty$ and $E(B_c) = E(B)$. This implies

$$\det \Delta(B_c) = \det \Delta(B) \ne 0.$$

Since $\Delta(B_c)$ and $\Delta(B)$ are strictly positive operators with $\Delta(B_c) \ge \Delta(B)$, we infer that $\det \Delta(B_c) = \det \Delta(B)$ if and only if $\Delta(B_c) = \Delta(B)$. Using the uniqueness part of Theorem 2.8, we get $B = B_c$, which completes the proof. □

THEOREM 2.18. *Let $[T_1' \ \cdots \ T_n']$, $T_i' \in B(\mathcal{H}')$, be a row contraction with minimal isometric dilation $\mathcal{V}'$ and let $A : F^2(H_n) \otimes \mathcal{E} \to \mathcal{H}'$ be such that $\|A\| < t$ and*

$$A(S_i \otimes I_{\mathcal{E}}) = T_i' A, \quad i = 1, \ldots, n.$$

If B_c is the central intertwining lifting of A with respect to $\mathcal{V}'$ and tolerance t, then the multi-Toeplitz operator $t^2 I - B_c^* B_c$ admits a square outer spectral factorization

$$t^2 I - B_c^* B_c = M_\varphi^* M_\varphi,$$

where $M_\varphi \in R_n^\infty \bar{\otimes} B(\mathcal{E})$ is an outer operator given by

$$M_\varphi := [I \otimes (P_{\mathcal{E}} \psi)^{1/2}] M_\psi^{-1},$$

and the symbol ψ of the multi-analytic operator M_ψ is given by $\psi := (t^2 I - A^* A)^{-1}|\mathcal{E}$.

Moreover, if $\dim \mathcal{E} < \infty$, then

$$E(B_c) = -\ln\det[P_{\mathcal{E}}(t^2 I - A^* A)^{-1}|\mathcal{E}]. \tag{2.62}$$

PROOF. Using the proof of Corollary 2.15 and Corollary 2.16, we deduce that

$$(I \otimes \Omega^*) Z = Z_{\mathcal{E}} = M_\varphi,$$

and M_φ is the outer spectral factor of $t^2 I - B_c^* B_c$ defined by (2.61). Recall that $\mathcal{D}_{A,t} \ominus \mathcal{N} = \mathcal{M} = D_{A,t}^{-1}\mathcal{E}$ and

$$\Omega := (D_{A,t}^{-1}|\mathcal{E})(P_{\mathcal{E}} D_{A,t}^{-2}|\mathcal{E})^{-1/2}$$

is a unitary operator from $\mathcal{E}$ onto $\mathcal{M}$. Hence, we have $P_{\mathcal{N}} D_{A,t}^{-1}|\mathcal{H} = 0$ and

$$\begin{aligned} M_\varphi D_{A,t}^{-2} h &= (I \otimes \Omega^*) Z D_{A,t}^{-2} h = \Omega^* P_{\mathcal{M}} D_{A,t}^{-1} h \\ &= \Omega^* D_{A,t}^{-1} h = (P_{\mathcal{E}} D_{A,t}^{-2}|\mathcal{E})^{1/2} h = 1 \otimes (P_{\mathcal{E}} \psi)^{1/2} h \end{aligned} \tag{2.63}$$

for any $h \in \mathcal{E}$. Applying Theorem 1.5 to the strictly positive multi-Toeplitz operator $T := t^2 I - A^*A$, we infer that M_ψ is an invertible multi-analytic operator. Now, note that

$$M_\varphi M_\psi(e_\alpha \otimes h) = (S_\alpha \otimes I_\mathcal{E})M_\varphi D_{A,t}^{-2} = e_\alpha \otimes (P_\mathcal{E}\psi)^{1/2} h$$

for any $h \in \mathcal{E}$ and $\alpha \in \mathbb{F}_n^+$. Hence,

$$M_\varphi = [I_{F^2(H_n)} \otimes (P_\mathcal{E}\Psi)^{1/2}]M_\psi^{-1}.$$

To complete the proof, note that relation (2.61) implies

$$\varphi(0)^*\varphi(0) = \Delta(A) = P_\mathcal{E}(t^2 I - A^*A)^{-1}|\mathcal{E}. \tag{2.64}$$

Since M_φ is outer, Theorem 1.12 implies

$$E(B_c) = e(M_\varphi^* M_\varphi) = \ln\det[\varphi(0)^*\varphi(0)]. \tag{2.65}$$

Combining relations (2.64) and (2.65), we get (2.62). The proof is complete. □

If B be an arbitrary intertwining lifting of A with $\|B\| < t$, we can obtain a result similar to that of Theorem 2.18 by applying Theorem 1.1 and Corollary 1.2 to the multi-Toeplitz operator $T := t^2 I - B^*B$.

Corollary 2.19. *Let $[T_1' \cdots T_n']$, $T_i' \in B(\mathcal{H}')$, be a row contraction with minimal isometric dilation $\mathcal{V}'$ and let $A : F^2(H_n) \otimes \mathcal{E} \to \mathcal{H}'$ be such that $\|A\| < t$ and*

$$A(S_i \otimes I_\mathcal{E}) = T_i' A, \quad i = 1, \ldots, n.$$

If B is an intertwining lifting of A with respect to $\mathcal{V}'$ and tolerance t, then the multi-Toeplitz operator $t^2 I - B^*B$ admits a square outer spectral factorization

$$t^2 I - B^*B = M_\chi^* M_\chi,$$

where $M_\chi \in R_n^\infty \bar{\otimes} B(\mathcal{E})$ is an outer operator given by

$$M_\chi := [I \otimes (P_\mathcal{E} g)^{1/2}]M_g^{-1},$$

and the symbol g of the multi-analytic operator M_g is given by $g := (t^2 I - B^*B)^{-1}|\mathcal{E}$.

Moreover, if $\dim \mathcal{E} < \infty$, then

$$E(B) = -\ln\det[P_\mathcal{E}(t^2 I - B^*B)^{-1}|\mathcal{E}].$$

In what follows, we show that, under a certain condition of stability, there is a maximal principle for the noncommutative commutant lifting theorem and a characterization of the central intertwining lifting $\tilde{B}_c$ of A with respect to non-minimal isometric liftings. The results will be very useful in the next sections.

Theorem 2.20. *Let $\mathcal{T} := [T_1 \cdots T_n]$, $T_i \in B(\mathcal{H})$, be a row isometry and $\mathcal{T}' := [T_1' \cdots T_n']$, $T_i' \in B(\mathcal{H}')$, be a row contraction. Let $\mathcal{W}' := [W_1' \cdots W_n']$, $V_i' \in B(\mathcal{G}')$, be an isometric lifting of $\mathcal{T}'$, and let $A \in B(\mathcal{H}, \mathcal{H}')$ be such that $\|A\| \le t$ and*

$$AT_i = T_i' A, \quad i = 1, \ldots, n.$$

If B is an intertwining lifting of A with respect to $\mathcal{W}'$ and tolerance $t > 0$, then

$$\Delta(B) \le \Delta(\tilde{B}_c) = \Delta(A), \tag{2.66}$$

where $\tilde{B}_c$ is the central intertwining lifting of A with tolerance $t > 0$. Assume that $[E_1^* \cdots E_n^*]$ is a C_0-row contraction, where $E_i := P_i W P_\mathcal{N} \in B(\mathcal{D}_{A,t})$, for any $i = 1, \ldots, n$. Then $\Delta(B) = \Delta(A)$ if and only if $B = \tilde{B}_c$.

In particular, if $\|A\| < t$ and $[T_1 \ \cdots \ T_n]$ is unitarily equivalent to an orthogonal shift $[S_1 \otimes I \ \cdots \ S_n \otimes I]$, then $\Delta(B) = \Delta(A)$ if and only if $B = \tilde{B}_c$.

PROOF. The inequality (2.66) was proved in Theorem 2.9. Assume now that the row operator $[E_1^* \ \cdots \ E_n^*]$ is a C_0-row contraction and $\Delta(B) = \Delta(A)$. Recall that the operator W_i' admits a reducing decomposition $W_i' = V_i' \oplus U_i'$, $i = 1, \dots, n$, of $\mathcal{G}' = \mathcal{K}' \oplus \mathcal{E}'$ such that $\mathcal{V}' := [V_1' \ \cdots \ V_n']$ is the minimal isometric dilation of $\mathcal{T}'$ on $\mathcal{K}'$, which can be identified with $\mathcal{H}' \oplus [F^2(H_n) \otimes \mathcal{D}']$. Since $P_{\mathcal{H}'}B = A$, the operator B has the matrix representation

$$B = \begin{bmatrix} A \\ \Lambda D_{A,t} \\ X D_{A,t} \end{bmatrix} : \mathcal{H} \to \mathcal{H}' \oplus [F^2(H_n) \otimes \mathcal{D}'] \oplus \mathcal{E}', \tag{2.67}$$

where the operator

$$\begin{bmatrix} \Lambda \\ X \end{bmatrix} : \mathcal{D}_{A,t} \to [F^2(H_n) \otimes \mathcal{D}'] \oplus \mathcal{E}'$$

is a contraction. According to Theorem 2.9, the equation $\Delta(B) = \Delta(A)$ implies $\begin{bmatrix} \Lambda \\ X \end{bmatrix} |\mathcal{M} = 0$, i.e., $\Lambda|\mathcal{M} = 0$ and $X|\mathcal{M} = 0$. According to Remark 2.5, we deduce that $B_c = \begin{bmatrix} A \\ \Lambda D_{A,t} \end{bmatrix}$ is the central intertwining lifting of A with respect to $\mathcal{V}'$ and tolerance t. Since $\begin{bmatrix} \Lambda \\ X \end{bmatrix}$ is a contraction, there exists another contraction $G : \mathcal{D}_\Lambda \to \mathcal{E}'$ such that $X = GD_\Lambda$. On the other hand, since $\Lambda|\mathcal{M} = 0$, we have $D_\Lambda|\mathcal{M} = I_\mathcal{M}$. From $X|\mathcal{M} = 0$, we deduce $G|\mathcal{M} = 0$. The matrix representation (2.67) becomes

$$B = \begin{bmatrix} A \\ \Lambda D_{A,t} \\ G D_\Lambda D_{A,t} \end{bmatrix}, \tag{2.68}$$

where Λ and G are contraction such that $\Lambda|\mathcal{M} = 0$ and $G|\mathcal{M} = 0$. Since $V_i' B_c = B_c T_i$ for any $i = 1, \dots, n$, we have

$$\begin{aligned} \left\| D_{B_c,t} \left(\sum_{i=1}^n T_i h_i \right) \right\|^2 &= \left\langle \sum_{i,j=1}^n T_i^* D_{B_c,t}^2 T_i h_i, h_j \right\rangle \\ &= \sum_{i,j=1}^n \langle \delta_{ij} D_{B_c,t}^2 h_i, h_j \rangle = \sum_{i=1}^n \|D_{B_c,t}^2 h_i\|^2. \end{aligned}$$

Hence, we deduce that there are some unique isometries $V_i \in B(\mathcal{D}_{B_c,t})$, $i = 1, \dots, n$, such that

$$V_i D_{B_c,t} = D_{B_c,t} V_i, \quad , i = 1, \dots, n, \tag{2.69}$$

and

$$\left\| \sum_{i=1}^n V_i D_{B_c,t} h_i \right\|^2 = \sum_{i=1}^n \|D_{B_c,t} h_i\|^2.$$

This proves that $V_1, \dots, V_n$ are isometries with orthogonal ranges. Since $[E_1^* \ \cdots \ E_n^*]$ is a C_0-row contraction, we have $Q = 0$ in Lemma 2.12. Therefore, Theorem 2.14 implies

$$t^2 I - B_c^* B_c = Z^* Z \quad \text{and} \quad Z T_i = (S_i \otimes I_\mathcal{M}) Z \tag{2.70}$$

for any $i = 1, \dots, n$, where $Z : \mathcal{H} \to F^2(H_n) \otimes \mathcal{M}$ has dense range and is defined as in Theorem 2.14. According to the factorization (2.70), there exists an isometric operator $U : \mathcal{D}_{B_c,t} \to F^2(H_n) \otimes \mathcal{M}$ such that

$$UD_{B_c,t} = Z. \tag{2.71}$$

Since Z has dense range, we infer that U is a unitary operator. Using relations (2.69), (2.70), and (2.71), we obtain

$$\begin{aligned} UV_iD_{B_c,t} &= UD_{B_c,t}T_i = ZT_i \\ &= (S_i \otimes I_{\mathcal{M}})Z = (S_i \otimes I_{\mathcal{M}})UD_{B_c,t}. \end{aligned}$$

Therefore,

$$UV_i = (S_i \otimes I_{\mathcal{M}})U, \quad i = 1, \dots, n.$$

On the other hand, since $B_c = \begin{bmatrix} A \\ \Lambda D_{A,t} \end{bmatrix}$ and Λ is a contraction, there is a unitary operator $U_\Lambda : \mathcal{D}_{B_c,t} \to \mathcal{D}_\Lambda$ such that

$$U_\Lambda D_{B_c,t} = D_\Lambda D_{A,t}. \tag{2.72}$$

Now, define the isometries $W_i \in B(\mathcal{D}_\Lambda)$ by setting

$$W_i := U_\Lambda V_i U_\Lambda^*, \quad i = 1, \dots, n. \tag{2.73}$$

Note that relations (2.72), (2.73), and (2.69) imply

$$W_i D_\Lambda D_{A,t} = D_\Lambda D_{A,t} T_i, \quad i = 1, \dots, n. \tag{2.74}$$

Indeed, we have

$$U_\Lambda V_i U_\Lambda^* D_\Lambda D_{A,t} = U_\Lambda V_i D_{B_c,t} = U_\Lambda D_{B_c,t} T_i = D_\Lambda D_{A,t} T_i$$

for any $i = 1, \dots, n$. Now, we prove that the subspace $\mathcal{M} := \mathcal{D}_{A,t} \ominus \bigvee_{i=1}^n D_{A,t} T_i \mathcal{H}$ is equal to $\bigcap_{i=1}^n \ker W_i^*$. According to the Wold type decomposition for sequences of isometries with orthogonal ranges (see [**38**]) and using (2.73), we have

$$\begin{aligned} \bigcap_{i=1}^n \ker W_i^* &= \mathcal{D}_\Lambda \ominus [W_1 \mathcal{D}_\Lambda \oplus \cdots \oplus W_1 \mathcal{D}_\Lambda] \\ &= \mathcal{D}_\Lambda \ominus [\oplus_{i=1}^n W_i D_\Lambda D_{A,t} \mathcal{H}] \\ &= \mathcal{D}_\Lambda \ominus [\oplus_{i=1}^n D_\Lambda D_{A,t} T_i \mathcal{H}] \\ &= \mathcal{D}_\Lambda \ominus D_\Lambda \mathcal{N}. \end{aligned}$$

Since $\Lambda|\mathcal{M} = 0$, we have $D_\Lambda|\mathcal{M} = I_\mathcal{M}$ and $\mathcal{M}$ is a reducing subspace for D_Λ. Hence, we deduce that

$$\mathcal{D}_\Lambda = \overline{D_\Lambda \mathcal{M}} \oplus \overline{D_\Lambda \mathcal{N}} = \mathcal{M} \oplus \overline{D_\Lambda \mathcal{N}}.$$

The previous computations show that $\bigcap_{i=1}^n \ker W_i^* = \mathcal{M}$. Now, since $W_i' B = BT_i$ and $W_i' = V_i' \oplus U_i'$ for any $i = 1, \dots, n$, we can take into account the matrix representation (2.68) and relation (2.74) to deduce

$$U_i' G D_\Lambda D_{A,t} = G D_\Lambda D_{A,t} T_i = G W_i D_\Lambda D_{A,t}$$

for any $i = 1, \dots, n$. Hence, we get $U_i' G = G W_i$, $i = 1, \dots, n$. Using the fact that $G|\mathcal{M} = 0$, we obtain

$$G W_\alpha h = U_\alpha' h = 0$$

for any $\alpha \in \mathbb{F}_n^+$ and $h \in \mathcal{M}$. Since $\bigvee_{\alpha\in\mathbb{F}_n^+} W_\alpha \mathcal{M} = \mathcal{D}_\Lambda$ and $G : \mathcal{D}_\Lambda \to \mathcal{E}'$, we get $G = 0$. Therefore, $B = \begin{bmatrix} B_c \\ 0 \end{bmatrix} = \tilde{B}_c$ is the central intertwining lifting of A. To prove that last part of the theorem, note that if $\|A\| < t$ and $[T_1 \ \cdots \ T_n]$ is an orthogonal shift, then according to Lemma 2.12, the operator $[E_1^* \ \cdots \ E_n^*]$ is a C_0-row contraction. Therefore, we can apply the first part of the theorem to complete the proof. □

The following result is now a consequence of the previous theorem. Since the proof is similar to that of Theorem 2.17, we shall omit it.

PROPOSITION 2.21. *Let $\mathcal{T}' := [T_1' \ \cdots \ T_n']$, $T_i' \in B(\mathcal{H}')$, be a row contraction, $\mathcal{W}' := [W_1' \ \cdots \ W_n']$ be an isometric lifting of $\mathcal{T}'$, and let $A \in B(F^2(H_n) \otimes \mathcal{E}, \mathcal{H}')$ be such that $\|A\| \leq t$ and*

$$A(S_i \otimes I_\mathcal{E}) = T_i' A, \quad i = 1, \ldots, n.$$

If B is an intertwining lifting of A with respect to $\mathcal{W}'$ and tolerance $t > 0$, then

$$E(A) = E(\tilde{B}_c) \geq E(B), \tag{2.75}$$

where $\tilde{B}_c$ is the central intertwining lifting of A with tolerance $t > 0$. Assume that $[E_1^* \ \cdots \ E_n^*]$ is a C_0-row contraction, where $E_i := P_i W P_\mathcal{N} \in B(\mathcal{D}_{A,t})$ for any $i = 1, \ldots, n$. If the entropy $E(\tilde{B}_c) > -\infty$, then $E(\tilde{B}_c) = E(B)$ if and only if $B = \tilde{B}_c$.

In particular, if $\|A\| < t$, then $E(\tilde{B}_c) = E(B)$ if and only if $B = \tilde{B}_c$.

We remark that if $\|A\| < t$, then the operator $\Delta(A) \in B(\mathcal{E})$ is invertible and $E(A) > -\infty$.

CHAPTER 3

Maximal Entropy Interpolation Problems in Several Variables

We obtain explicit forms for the maximal entropy solution (as well as its entropy) of the Sarason [**60**], Carathéodory-Schur [**12**], [**61**], and Nevanlinna-Pick [**35**] type interpolation problems for the noncommutative (resp. commutative) analytic Toeplitz algebra F_n^∞ (resp. W_n^∞) and their tensor products with $B(\mathcal{H},\mathcal{K})$. Moreover, under certain conditions, we also find explicit forms for the corresponding classical optimization problems, in our multivariable noncommutative (resp. commutative) setting. In particular, we provide explicit forms for the maximal entropy solutions of several interpolation problems on the unit ball of $\mathbb{C}^n$. Finnaly, we apply our permanence principle to the Nevanlinna-Pick interpolation problem on the unit ball.

3.1. Maximal entropy solution for the Sarason interpolation problem for analytic Toeplitz algebras

Given $R \in R_n^\infty \bar{\otimes} B(\mathcal{E},\mathcal{E}')$ with $\dim \mathcal{E} < \infty$, and $\Phi \in R_n^\infty \bar{\otimes} B(\mathcal{E}_1,\mathcal{E}')$ an inner operator, we consider the following Sarason (see [**60**]) type left interpolation problem with tolerance $t > 0$ for $R_n^\infty \bar{\otimes} B(\mathcal{E},\mathcal{E}')$:

Find $\Psi \in R_n^\infty \bar{\otimes} B(\mathcal{E},\mathcal{E}')$ *with maximal entropy such that*

$$\|\Psi\| \leq t \quad \textit{and} \quad \Psi = R + \Phi G, \tag{3.1}$$

where $G \in R_n^\infty \bar{\otimes} B(\mathcal{E},\mathcal{E}_1)$.

In this section, we find the solution to this problem and prove a permanence principle for it (see Theorem 3.1 and Theorem 3.2). Under a certain condition, we obtain explicit forms for the maximal entropy solution of the Sarason interpolation problem for analytic Toeplitz algebras in our multivariable setting (see Theorem 3.3 and Corollary 3.4). We also find an explicit form for the unique solution of the corresponding Sarason optimization problem (see Theorem 3.5 and Theorem 3.6) in our setting.

We recall (from Section 1.2) that the entropy of a multi-analytic bounded operator $\Psi \in R_n^\infty \bar{\otimes} B(\mathcal{E},\mathcal{E}')$ with $\|\Psi\| \leq t$ is defined by $E(\Psi) := \ln \det \Delta(\Psi)$, where

$$\langle \Delta(\Psi)h, h\rangle := \inf \left\langle (t^2 I - \Psi^*\Psi)(h-p), h-p\right\rangle, \quad h \in \mathcal{E},$$

where the infimum is taken over all polynomials $p \in F^2(H_n) \otimes \mathcal{E}$ with $p(0) = 0$.

THEOREM 3.1. *Let* $R \in R_n^\infty \bar{\otimes} B(\mathcal{E},\mathcal{E}')$ *with* $\dim \mathcal{E} < \infty$, *and* $\Phi \in R_n^\infty \bar{\otimes} B(\mathcal{E}_1,\mathcal{E}')$ *be an inner operator. Let* $A : F^2(H_n) \otimes \mathcal{E} \to \mathcal{H}'$ *be defined by* $A := P_{\mathcal{H}'} R$, *where*

$$\mathcal{H}' := [F^2(H_n) \otimes \mathcal{E}'] \ominus \Phi[F^2(H_n) \otimes \mathcal{E}_1]. \tag{3.2}$$

Then there is a solution for the interpolation problem (3.1) *if and only if* $\|A\| \leq t$. *Moreover, the central intertwining lifting* $\Psi_{\max}$ *of* A *with respect to* $\{S_i \otimes I_{\mathcal{E}}\}_{i=1}^n$ *and* $\{S_i \otimes I_{\mathcal{E}'}\}_{i=1}^n$ *is the maximal entropy solution for the problem* (3.1).

PROOF. First, note that if Ψ is a solution to the interpolation problem (3.1), then we have $P_{\mathcal{H}'}\Psi = P_{\mathcal{H}'}R = A$ and $\|A\| \leq t$. Conversely, let $T_i \in B(\mathcal{H}')$, $i = 1, \ldots, n$, be defined by $T_i' := P_{\mathcal{H}'}(S_i \otimes I_{\mathcal{E}'})|\mathcal{H}'$ and note that

$$A(S_i \otimes I_{\mathcal{E}}) = T_i' A, \quad i = 1, \ldots, n.$$

Since $[S_1 \otimes I_{\mathcal{E}'} \ \cdots \ S_n \otimes I_{\mathcal{E}'}]$ is an isometric dilation of $[T_1' \ \cdots \ T_n']$, we can apply Theorem 2.4 and the remarks following Theorem 2.5 to this setting. Therefore, the central intertwining lifting $\tilde{B}_c$ of A with tolerance t satisfies the conditions $P_{\mathcal{H}'}\tilde{B}_c = A$, $\|\tilde{B}_c\| \leq t$, and

$$\tilde{B}_c(S_i \otimes I_{\mathcal{E}}) = (S_i \otimes I_{\mathcal{E}'})\tilde{B}_c, \quad i = 1, \ldots, n.$$

Hence, $\tilde{B}_c = \Psi_{\max}$ with $\Psi_{\max} \in R_n^\infty \bar{\otimes} B(\mathcal{E}, \mathcal{E}')$. Since $P_{\mathcal{H}'}(\Psi_{\max} - R) = 0$, according to [**53**], we have $\Psi_{\max} - R \in \Phi(R_n^\infty \bar{\otimes} B(\mathcal{E}, \mathcal{E}'))$. Hence, $\Psi_{\max} = R + \Phi G$ for some $G \in R_n^\infty \bar{\otimes} B(\mathcal{E}, \mathcal{E}_1)$. On the other hand, since $\Psi_{\max}$ is the central intertwining lifting of A, Theorem 2.20 and Proposition 2.21 show that it is also the maximal entropy solution to the interpolation problem (3.1). The proof is complete. □

Let $\mathcal{K}' \subseteq \mathcal{E}'$ be an invariant subspace under each operator $S_i^* \otimes I_{\mathcal{E}'}$, $i = 1, \ldots, n$, such that $\mathcal{H}' \subseteq \mathcal{K}'$, where $\mathcal{H}'$ is given by relation (3.2). According to Theorem 2.2 from [**39**], there exists an inner operator $\Psi \in R_n^\infty \bar{\otimes} B(\mathcal{E}_2, \mathcal{E}')$ such that

$$\mathcal{K}' = [F^2(H_n) \otimes \mathcal{E}'] \ominus \Psi[F^2(H_n) \otimes \mathcal{E}_2]. \tag{3.3}$$

Since $\mathcal{K}' \supseteq \mathcal{H}'$, we have

$$\Psi[F^2(H_n) \otimes \mathcal{E}_2] \subset \Phi[F^2(H_n) \otimes \mathcal{E}_1].$$

Hence, as in the proof of Theorem 3.7 from [**53**], one can show that there is $\Phi_1 \in R_n^\infty \bar{\otimes} B(\mathcal{E}_2, \mathcal{E}_1)$ such that $\Psi = \Phi\Phi_1$. Conversely, if $\Psi = \Phi\Phi_1$, then the subspace $\mathcal{K}'$ defined by (3.3) defines an invariant subspace under each operator $S_i^* \otimes I_{\mathcal{E}'}$, $i = 1, \ldots, n$.

Let $\Phi \in R_n^\infty \bar{\otimes} B(\mathcal{E}_1, \mathcal{E}')$ and $\Phi_1 \in R_n^\infty \bar{\otimes} B(\mathcal{E}_2, \mathcal{E}_1)$ be inner operators, and let $\Psi_{\max} \in R_n^\infty \bar{\otimes} B(\mathcal{E}, \mathcal{E}')$ be the maximal entropy solution of the problem (3.1). Consider the following interpolation problem:

Find $\Gamma \in R_n^\infty \bar{\otimes} B(\mathcal{E}, \mathcal{E}')$ *with maximal entropy such that*

$$\|\Gamma\| \leq t \quad \text{and} \quad \Gamma = \Psi_{\max} + \Phi\Phi_1 G, \tag{3.4}$$

where $G \in R_n^\infty \bar{\otimes} B(\mathcal{E}, \mathcal{E}_2)$.

We can prove the following permanence principle for the interpolation problem (3.1).

THEOREM 3.2. *Let* $R \in R_n^\infty \bar{\otimes} B(\mathcal{E}, \mathcal{E}')$ *with* $\dim \mathcal{E} < \infty$, *and* $\Phi \in R_n^\infty \bar{\otimes} B(\mathcal{E}_1, \mathcal{E}')$ *be a pure inner operator. Let* $\Psi_{\max}$ *be the maximal entropy solution for the interpolation problem* (3.1). *If* $\Phi_1 \in R_n^\infty \bar{\otimes} B(\mathcal{E}_2, \mathcal{E}_1)$ *is an inner operator, then* $\Psi_{\max}$ *is also the maximal entropy solution for the interpolation problem* (3.4).

PROOF. Let $\Psi_{\max}$ be the central intertwining lifting of the operator $A := P_{\mathcal{H}'}R$. Then according to Theorem 3.1, $\Psi_{\max}$ is a maximal entropy solution for the interpolation problem (3.1). Note that since Φ is pure, i.e., $\|P_{\mathcal{E}'}\Phi h\| < \|h\|$, $h \in \mathcal{E}_1$, the results from [**39**] imply that $[S_1 \otimes I_{\mathcal{E}'} \ \cdots \ S_n \otimes I_{\mathcal{E}'}]$ is the minimal isometric dilation of the row contraction $[T_1' \ \cdots \ T_n']$. Set

$$\mathcal{K}' = [F^2(H_n) \otimes \mathcal{E}'] \ominus \Phi\Phi_1[F^2(H_n) \otimes \mathcal{E}_2]$$

and let $Y_{\mathcal{K}'} := P_{\mathcal{K}'}\Psi_{\max}$ be the corresponding central partial intertwining lifting of A. The permanence principle of Theorem 2.10 implies that $\Psi_{\max}$ is also the central intertwining lifting of $Y_{\mathcal{K}'}$. Applying Theorem 3.1 to the operator $Y_{\mathcal{K}'}$ instead of A, we conclude that $\Psi_{\max}$ is also the maximal entropy solution of the interpolation problem (3.4). □

We remark that the previous theorem will be used in Section 3.4 to obtain a permanence principle for the Nevanlinna-Pick interpolation problem on the unit ball of $\mathbb{C}^n$.

Now, using the results of the previous sections, we can obtain explicit forms for the maximal entropy solution of the Sarason type interpolation problem for $R_n^\infty \bar{\otimes} B(\mathcal{E}, \mathcal{Y})$.

THEOREM 3.3. *Let $\mathcal{H}' \subset F^2(H_n) \otimes \mathcal{Y}$ be an invariant subspace under each operator $S_i^* \otimes I_{\mathcal{Y}}$, $i = 1, \ldots, n$, and let $A : F^2(H_n) \otimes \mathcal{E} \to \mathcal{H}'$ be such that $\|A\| < t$ and*

$$A(S_i \otimes I_{\mathcal{E}}) = [P_{\mathcal{H}'}(S_i \otimes I_{\mathcal{Y}})|\mathcal{H}']A, \quad i = 1, \ldots, n.$$

Let $\psi : \mathcal{E} \to F^2(H_n) \otimes \mathcal{E}$ and $\theta : \mathcal{E} \to F^2(H_n) \otimes \mathcal{Y}$ be operators defined by

$$\text{(3.5)} \qquad \psi h := (t^2 I - A^*A)^{-1}(1 \otimes h) \quad \text{and} \quad \theta h := A(t^2 I - A^*A)^{-1}(1 \otimes h),$$

for any $h \in \mathcal{E}$. Then M_ψ, M_θ are multi-analytic operators, with M_ψ invertible and

$$M_\psi^{-1}(1 \otimes h) = \sum_{\sigma \in \mathbb{F}_n^+} e_\sigma \otimes \Delta(A) P_{\mathcal{E}}(X_\sigma^A)^* h, \qquad h \in \mathcal{E}.$$

The central intertwining lifting $\tilde{B}_c$ of A with respect to $\{S_i \otimes I_{\mathcal{E}}\}_{i=1}^n$ and $\{S_i \otimes I_{\mathcal{Y}}\}_{i=1}^n$ is equal to $M_\theta M_\psi^{-1}$ and the multi-Toeplitz operator $t^2 I - \tilde{B}_c^* \tilde{B}_c$ admits a square outer spectral factorization

$$t^2 I - \tilde{B}_c^* \tilde{B}_c = M_\chi^* M_\chi,$$

where $M_\chi \in R_n^\infty \bar{\otimes} B(\mathcal{E})$ is an outer operator given by

$$M_\chi := [I \otimes (P_{\mathcal{E}}\psi)^{1/2}] M_\psi^{-1}.$$

Moreover, if $\dim \mathcal{E} < \infty$, then $M_\theta M_\psi^{-1}$ is the maximal entropy lifting of A and

$$E(M_\theta M_\psi^{-1}) = -\ln \det[P_{\mathcal{E}}(t^2 I - A^*A)^{-1}|\mathcal{E}].$$

PROOF. Note that $[S_1 \otimes I_{\mathcal{Y}} \ \cdots \ S_n \otimes I_{\mathcal{Y}}]$ is an isometric lifting of the row-contraction

$$C := [P_{\mathcal{H}'}(S_1 \otimes I_{\mathcal{Y}})|\mathcal{H}' \ \cdots \ P_{\mathcal{H}'}(S_n \otimes I_{\mathcal{Y}})|\mathcal{H}'].$$

According to the remarks following Theorem 2.5, there is a unique isometry

$$\Phi : \mathcal{K}' := \mathcal{H}' \oplus [F^2(H_n) \otimes \mathcal{D}'] \to F^2(H_n) \otimes \mathcal{Y},$$

such that $(S_i\otimes I_{\mathcal{Y}})\Phi = \Phi V_i'$, for any $i=1,\dots,n$, and $\Phi|\mathcal{H}'=\mathcal{H}'$, where $[V_1' \cdots V_n']$ is the minimal isometric dilation of $\mathcal{C}$ on $\mathcal{K}'$. The operator $\tilde{B}_c := \Phi B_c$ is the central intertwining dilation of A with respect to $[S_1\otimes I_{\mathcal{Y}} \ \cdots \ S_n\otimes I_{\mathcal{Y}}]$. Since

$$(S_i\otimes I_{\mathcal{Y}})\tilde{B}_c = \tilde{B}_c(S_i\otimes I_{\mathcal{E}}),\ i=1,\dots,n,$$

we infer that $\tilde{B}_c = M_f$ for some multi-analytic operator $M_f\in R_n^\infty\bar{\otimes}B(\mathcal{E},\mathcal{Y})$. On the other hand we have $\|M_f\| = \|B_c\|\le t$. According to Theorem 2.18, M_ψ is an invertible multi-analytic operator on $F^2(H_n)\otimes\mathcal{E}$. Using Lemma 2.11, we get

$$\begin{aligned}\tilde{B}_c\psi h &= \Phi B_c D_{A,t}^{-2}(1\otimes h) = \Phi A D_{A,t}^{-2}(1\otimes h)\\ &= A D_{A,t}^{-2}(1\otimes h) = \theta h\end{aligned}$$

for any $h\in\mathcal{E}$. Therefore $\tilde{B}_c M_\psi = M_\theta$, and $M_\theta\in R_n^\infty\bar{\otimes}B(\mathcal{E})$. Now, using Corollary 2.16 and Theorem 2.18 we can complete the proof of the theorem. □

Let $F_s^2(H_n)\subset F^2(H_n)$ be the symmetric Fock space and let us define $W_n^\infty := P_{F_s^2(H_n)}F_n^\infty|F_s^2(H_n)$. We recall that W_n^∞ is the WOT-closed algebra generated by

$$B_i := P_{F_s^2(H_n)}S_i|F_s^2(H_n),\quad i=1,\dots,n,$$

and the identity. Let $\mathcal{H}_s\subseteq F_s^2(H_n)\otimes\mathcal{Y}$ be an invariant subspace under each operator $B_i^*\otimes I_{\mathcal{Y}},\ \ i=1,\dots,n$, and let $C: F_s^2(H_n)\otimes\mathcal{E}\to\mathcal{H}_s$ be an operator satisfying $\|C\|\le t$ and

$$C(B_i\otimes I_{\mathcal{E}}) = [P_{\mathcal{H}_s}(B_i\otimes I_{\mathcal{Y}})|\mathcal{H}_s]C,\quad i=1,\dots,n. \tag{3.6}$$

Since $F_s^2(H_n)$ is an invariant subspace under each $S_i^*,\ \ i=1,\dots,n$, it is easy to see that $\mathcal{H}_s$ is also invariant under $S_i^*\otimes I_{\mathcal{Y}},\ \ i=1,\dots,n$. Setting

$$A := C(P_{F_s^2(H_n)}\otimes I_{\mathcal{E}}) : F^2(H_n)\otimes\mathcal{E}\to F^2(H_n)\otimes\mathcal{Y},$$

relation (3.6) implies

$$A(S_i\otimes I_{\mathcal{E}}) = [P_{\mathcal{H}_s}(S_i\otimes I_{\mathcal{Y}})|\mathcal{H}_s]A,\quad i=1,\dots,n.$$

Let $\tilde{B}_c$ be the central intertwining lifting of A with respect to $\{S_i\otimes I_{\mathcal{E}}\}_{i=1}^n$ and $\{S_i\otimes I_{\mathcal{Y}}\}_{i=1}^n$, i.e.,

$$\tilde{B}_c\in R_n^\infty\bar{\otimes}B(\mathcal{E},\mathcal{Y}),\quad P_{\mathcal{H}_s}\tilde{B}_c = A,\quad \text{and}\quad \|\tilde{B}_c\|\le t.$$

Define

$$\tilde{C}_c := (P_{F_s^2(H_n)}\otimes I_{\mathcal{Y}})\tilde{B}_c|F_s^2(H_n)\otimes\mathcal{E}\in W_n^\infty\bar{\otimes}B(\mathcal{E},\mathcal{Y})$$

and note that $P_{\mathcal{H}_s}\tilde{C}_c = C$ and $\|\tilde{C}_c\|\le t$. We call $\tilde{C}_c$ the central intertwining lifting of C with respect to $\{B_i\otimes I_{\mathcal{E}}\}_{i=1}^n$ and $\{B_i\otimes I_{\mathcal{Y}}\}_{i=1}^n$, and tolerance t. Note that $\tilde{C}_c$ is a solution of the Sarason interpolation problem for the operator space $W_n^\infty\bar{\otimes}B(\mathcal{E},\mathcal{Y})$.

If $\|C\|<t$, then one can apply Theorem 3.3 to the bounded operator $A := C(P_{F_s^2(H_n)}\otimes I_{\mathcal{E}})$ and deduce the following multivariable commutative version for the Sarason interpolation problem.

Corollary 3.4. *Let $\mathcal{H}_s\subseteq F_s^2(H_n)\otimes\mathcal{Y}$ be an invariant subspace under each operator $B_i^*\otimes I_{\mathcal{Y}},\ \ i=1,\dots,n$, and let $C: F_s^2(H_n)\otimes\mathcal{E}\to\mathcal{H}_s$ be such that $\|C\|<t$ and*

$$C(B_i\otimes I_{\mathcal{E}}) = [P_{\mathcal{H}_s}(B_i\otimes I_{\mathcal{Y}})|\mathcal{H}_s]C,\quad i=1,\dots,n. \tag{3.7}$$

Let $\tilde{C}_c \in W_n^\infty \bar{\otimes} B(\mathcal{E}, \mathcal{Y})$ be defined by

$$\tilde{C}_c := (P_{F_s^2(H_n)} \otimes I_\mathcal{Y}) M_\theta M_\psi^{-1} | F_s^2(H_n) \otimes \mathcal{E},$$

where θ, ψ are defined by relation (3.5) and $A := C(P_{F_s^2(H_n)} \otimes I_\mathcal{E})$. Then $\tilde{C}_c$ is an intertwining lifting of C, i.e.,

$$P_{\mathcal{H}_s} \tilde{C}_c = C \quad \text{and} \quad \|\tilde{C}_c\| \leq t.$$

In what follows, we show that, under certain conditions on the operator A, there is an explicit form of the unique intertwining lifting B of A such that $\|A\| = \|B\|$. We recall that an operator $T \in B(\mathcal{X}, \mathcal{Y})$ attains its norm if there is a vector $x \in \mathcal{X}$ of norm one such that $\|Tx\| = \|T\|$.

THEOREM 3.5. *Let $\mathcal{H}' \subset F^2(H_n) \otimes \mathcal{Y}$ be an invariant subspace under each operator $S_i^* \otimes I_\mathcal{Y}$, $i = 1, \ldots, n$, and let $A : F^2(H_n) \to \mathcal{H}'$ be a contraction with $\|A\| = 1$ and which attains its norm. If*

$$AS_i = [P_{\mathcal{H}'}(S_i \otimes I_\mathcal{Y}) | \mathcal{H}'] A, \quad i = 1, \ldots, n,$$

then there is a unique intertwining lifting $M_\varphi \in R_n^\infty \bar{\otimes} B(\mathbb{C}, \mathcal{Y})$ of A such that $\|A\| = \|M_\varphi\|$. Moreover, M_φ is an inner operator uniquely defined by the equation

$$M_\varphi g = Ag \tag{3.8}$$

and g is a vector in $F^2(H_n)$ where A attains its norm.

PROOF. Let $g = \chi f$ be the inner-outer factorization of g, where $\chi \in F_n^\infty$ is inner and $f \in F^2(H_n)$ is outer. Since $AS_i = T_i' A$, $i = 1, \ldots, n$, where

$$T_i' := P_{\mathcal{H}'}(S_i \otimes I_\mathcal{Y}) | \mathcal{H}', \quad i = 1, \ldots, n,$$

and $[T_1' \cdots T_n']$ is a C_0-row contraction, we can use the F_n^∞-functional calculus for row contractions (see [**43**]) and obtain

$$\begin{aligned} \|A\| &= \|Ag\| = \|A\chi(S_1, \ldots, S_n) f\| \\ &= \|\chi(T_1', \ldots, T_n') A f\| \leq \|Af\| \\ &\leq \|A\| \|f\| = \|A\|. \end{aligned}$$

Here, we took into account that $\|f\|_2 = 1$ and $\|\chi\| = 1$. Therefore, $\|A\| = \|Af\|$. By the noncommutative commutant lifting theorem, there is an intertwining lifting B of A such that

$$BS_i = (S_i \otimes I_\mathcal{Y}) B, \quad i = 1, \ldots, n,$$

and $\|A\| = \|B\|$. According to [**44**], there is $M_\varphi \in R_n^\infty \bar{\otimes} B(\mathbb{C}, \mathcal{Y})$ such that $M_\varphi = B$. Since $P_\mathcal{H}' M_\varphi = A$, we have

$$\begin{aligned} \|A\| &= \|Af\| = \|P_\mathcal{H}' M_\varphi f\| \\ &\leq \|M_\varphi f\| \leq \|M_\varphi\| = \|A\|, \end{aligned}$$

which implies $Af = P_\mathcal{H}' M_\varphi f = M_\varphi f$. Since M_φ is a multi-analytic operator, we deduce that

$$M_\varphi \left(\sum_{|\alpha| \leq m} a_\alpha S_\alpha f \right) = \sum_{|\alpha| \leq m} a_\alpha (S_\alpha \otimes I_\mathcal{Y}) A f, \quad a_\alpha \in \mathbb{C}.$$

Note that, since $A^*Af = f$, we have

$$\begin{aligned}
\left\| \sum_{|\alpha|\leq m} a_\alpha (S_\alpha \otimes I_\mathcal{Y}) Af \right\|^2 &= \sum_{\alpha>\beta,\ |\alpha|,|\beta|\leq m} \langle a_{\alpha\backslash\beta}(S_{\alpha\backslash\beta}\otimes I_\mathcal{Y})Af, Af\rangle \\
&\quad + \sum_{\alpha<\beta,\ |\alpha|,|\beta|\leq m} \langle Af, a_{\beta\backslash\alpha}(S_{\beta\backslash\alpha}\otimes I_\mathcal{Y})Af\rangle \\
&= \sum_{\alpha>\beta,\ |\alpha|,|\beta|\leq m} \left\langle a_{\alpha\backslash\beta} T'_{\alpha\backslash\beta} Af, Af\right\rangle \\
&\quad + \sum_{\alpha<\beta,\ |\alpha|,|\beta|\leq m} \left\langle Af, a_{\beta\backslash\alpha} T'_{\beta\backslash\alpha} Af\right\rangle \\
&= \sum_{\alpha>\beta,\ |\alpha|,|\beta|\leq m} \langle a_{\alpha\backslash\beta} A(S_{\alpha\backslash\beta}\otimes I_\mathcal{Y})f, Af\rangle \\
&\quad + \sum_{\alpha<\beta,\ |\alpha|,|\beta|\leq m} \langle Af, a_{\beta\backslash\alpha} A(S_{\beta\backslash\alpha}\otimes I_\mathcal{Y})f\rangle \\
&= \sum_{\alpha>\beta,\ |\alpha|,|\beta|\leq m} \langle a_{\alpha\backslash\beta}(S_{\alpha\backslash\beta}\otimes I_\mathcal{Y})f, f\rangle \\
&\quad + \sum_{\alpha<\beta,\ |\alpha|,|\beta|\leq m} \langle f, a_{\beta\backslash\alpha}(S_{\beta\backslash\alpha}\otimes I_\mathcal{Y})f\rangle \\
&= \left\| \sum_{|\alpha|\leq m} a_\alpha S_\alpha f \right\|^2 .
\end{aligned}$$

Since f is outer, we infer that M_φ is an inner operator.

Assume now that there is another $M_{\varphi_1} \in R_n^\infty \bar{\otimes} B(\mathbb{C}, \mathcal{Y})$ such that $M_{\varphi_1} g = Ag$. Then we have $M_\varphi g = M_{\varphi_1} g$. Using the inner-outer factorization $g = \chi f$, we get $M_\varphi f = M_{\varphi_1} f$. Since f is outer, we have $M_\varphi = M_{\varphi_1}$. Therefore, $\varphi = \varphi_1$ and the proof is complete. □

According to the proof of Theorem 3.5, one can assume that $\|Af\| = 1$ and $M_\varphi f = Af$ for some outer vector f in $F^2(H_n)$. In this case, there is a sequence of polynomials $p_k \in F^2(H_n)$ such that

$$\|p_k(S_1, \ldots, S_n)f - 1\| \to 0, \qquad \text{as } k \to \infty.$$

Now, it is clear that

$$\varphi = \lim_{k\to\infty} p_k(S_1 \otimes I_\mathcal{Y}, \ldots, S_n \otimes I_\mathcal{Y})Af.$$

Note that if $\ker(I - A^*A)$ is one dimensional, then any vector where A attains its norm is outer.

Our commutative version of Theorem 3.5 is the following.

THEOREM 3.6. *Let $\mathcal{H}_s \subset F_s^2(H_n) \otimes \mathcal{Y}$ be an invariant subspace under each operator $B_i^* \otimes I_\mathcal{Y}$, $i = 1, \ldots, n$, and let $C : F_s^2(H_n) \to \mathcal{H}_s$ be a contraction which attains its norm $\|C\| = 1$ and*

$$CB_i = [P_{\mathcal{H}_s}(B_i \otimes I_\mathcal{Y})|\mathcal{H}_s]C, \quad i = 1, \ldots, n. \tag{3.9}$$

Then there exists $G \in W_n^\infty \bar{\otimes} B(\mathbb{C}, \mathcal{Y})$ *such that* $P_{\mathcal{H}_s} G = C$ *and* $\|G\| = \|C\|$. *Moreover, the operator* G *is given by the equation*

$$G = \frac{Cg}{g}, \tag{3.10}$$

where g *is any vector in* $F_s^2(H_n)$ *where* C *attains its norm.*

PROOF. Let $A := CP_{F_s^2(H_n)} : F^2(H_n) \to F^2(H_n) \otimes \mathcal{Y}$, and note that

$$\|A\| = \|C\| = \|Cg\| = \|Ag\| = 1.$$

Since $F_s^2(H_n)$ is an invariant subspace under each operator B_i^*, $i = 1, \dots, n$, relation (3.9) implies

$$AS_i = [P_{\mathcal{H}_s}(S_i \otimes I_{\mathcal{Y}})|\mathcal{H}_s]A, \quad i = 1, \dots, n.$$

According to Theorem 3.5, there exists $M_\varphi \in R_n^\infty \bar{\otimes} B(\mathbb{C}, \mathcal{Y})$ such that

$$P_{\mathcal{H}_s} M_\varphi = A, \ \|A\| = \|M_\varphi\|, \ \text{ and } \ M_\varphi g = Ag.$$

Hence, we infer that

$$(P_{F_s^2(H_n)\otimes \mathcal{Y}} M_\varphi | F_s^2(H_n))g = Cg.$$

Setting $G := P_{F_s^2(H_n)\otimes \mathcal{Y}} M_\varphi | F_s^2(H_n)$ and using the identification of W_n^∞ with the algebra of analytic multipliers of $F_s^2(H_n)$ (see [**6**]), it is clear that $G \in W_n^\infty \bar{\otimes} B(\mathbb{C}, \mathcal{Y})$ has the required properties. The proof is complete. □

We remark that one can obtain versions of Theorem 3.5 and Theorem 3.6 when $\|A\| = t$ and $t > 0$.

3.2. Maximal entropy solution for the Carathéodory-Schur interpolation problem for analytic Toeplitz algebras

Let $q := \sum_{|\alpha| \le m-1} R_\alpha \otimes A_\alpha$, $A_\alpha \in B(\mathcal{E}_1, \mathcal{E}_2)$, be a polynomial in $R_n^\infty \bar{\otimes} B(\mathcal{E}_1, \mathcal{E}_2)$ and let

$$d_\infty := \inf \left\{ \|G\| : \ G \in R_n^\infty \bar{\otimes} B(\mathcal{E}_1, \mathcal{E}_2) \text{ and } G_\alpha = A_\alpha, \text{ if } |\alpha| \le m-1 \right\}, \tag{3.11}$$

where $\{G_\alpha\}_{\alpha \in \mathbb{F}_n^+}$ are the Fourier coefficients of G. The Carathéodory-Schur (see [**12**], [**61**]) optimization problem (3.11) is to find $G \in R_n^\infty \bar{\otimes} B(\mathcal{E}_1, \mathcal{E}_2)$ with the smallest norm subject to the constraint

$$G_\alpha = A_\alpha \quad \text{if } \alpha \in \mathbb{F}_n^+, \ |\alpha| \le m-1. \tag{3.12}$$

The Carathéodory-Schur interpolation problem with tolerance $t > d_\infty$ is to find G with $\|G\| \le t$ subject to the the same constraint. In [**44**], we proved that $d_\infty = \|A\|$, where $A := P_{\mathcal{P}_{m-1} \otimes \mathcal{E}_2} q$ and $\mathcal{P}_{m-1}$ denotes the set of all polynomials in $F^2(H_n)$ of degree $\le m-1$.

In what follows, we provide an explicit form of the maximal entropy solution for the Carathéodory-Schur interpolation problem with tolerance $t > d_\infty$. For each $j = 1, 2$, define the Fourier transform $\mathcal{F}_j : \ell^2(\mathbb{F}_n^+) \otimes \mathcal{E}_j \to F^2(H_n) \otimes \mathcal{E}_j$ by setting

$$\mathcal{F}_j[(y_\alpha)_{\alpha \in \mathbb{F}_n^+}] = \sum_{\alpha \in \mathbb{F}_n^+} e_\alpha \otimes y_\alpha \tag{3.13}$$

for any $(y_\alpha)_{\alpha \in \mathbb{F}_n^+} \in \ell^2(\mathbb{F}_n^+) \otimes \mathcal{E}_j$.

THEOREM 3.7. *Let* $q := \sum_{|\alpha|\leq m-1} R_\alpha \otimes A_\alpha$*,* $A_\alpha \in B(\mathcal{E}_1, \mathcal{E}_2)$*, be a polynomial in* $R_n^\infty \bar{\otimes} B(\mathcal{E}_1, \mathcal{E}_2)$*, and let* $M := [A_{\alpha,\beta}]_{|\alpha|,|\beta|\leq m-1}$ *be the operator matrix defined by*

$$A_{\alpha,\beta} := \begin{cases} A_{\alpha\backslash\beta}, & \text{if } \alpha \geq \beta \\ 0, & \text{otherwise.} \end{cases}$$

Let $t > d_\infty = \|M\|$ *and define the operators* $\psi : \mathcal{E}_1 \to F^2(H_n) \otimes \mathcal{E}_1$ *and* $\theta : \mathcal{E}_1 \to F^2(H_n) \otimes \mathcal{E}_2$ *by*

$$\psi := \mathcal{F}_1(t^2 I - M^*M)^{-1} \begin{bmatrix} I_{\mathcal{E}_1} \\ 0 \\ \vdots \\ 0 \end{bmatrix} \quad \text{and} \quad \theta := \mathcal{F}_2 M (t^2 I - M^*M)^{-1} \begin{bmatrix} I_{\mathcal{E}_1} \\ 0 \\ \vdots \\ 0 \end{bmatrix}, \tag{3.14}$$

where $\mathcal{F}_1, \mathcal{F}_2$ *are the Fourier transforms defined by relation* (3.13)*. Then the operator*

$$M_\varphi := M_\theta M_\psi^{-1} \in R_n^\infty \bar{\otimes} B(\mathcal{E}_1, \mathcal{E}_2)$$

is the central solution for the Carathéodory-Schur interpolation problem with tolerance t*. Moreover, if* $\dim \mathcal{E}_1 < \infty$*, then* M_φ *is the maximal entropy solution.*

PROOF. Let $\mathcal{H}' := \mathcal{P}_{m-1} \otimes \mathcal{E}_2$ and $T_i' \in B(\mathcal{H}')$ be defined by $T_i' := P_{\mathcal{H}'}(S_i \otimes I_{\mathcal{E}_2})|\mathcal{H}'$, $i = 1, \ldots, n$. Setting

$$A := P_{\mathcal{H}'} q : F^2(H_n) \otimes \mathcal{E}_1 \to \mathcal{H}' \subset F^2(H_n) \otimes \mathcal{E}_2,$$

we can apply Theorem 3.3 to A and find the central intertwining lifting $\tilde{B}_c$ of A given by $\tilde{B}_c = M_\theta M_\psi^{-1}$ in $R_n^\infty \bar{\otimes} B(\mathcal{E}_1, \mathcal{E}_2)$, where θ, ψ are defined as in (3.5). Since $A\mathcal{F}_1 = \mathcal{F}_2[M\ 0]$, we have

$$\begin{aligned} \theta h &= A(t^2 I - A^*A)^{-1}(1 \otimes h) \\ &= A\mathcal{F}_1\mathcal{F}_1^{-1}(t^2 I - A^*A)^{-1}\mathcal{F}_1\mathcal{F}_1^{-1}(1 \otimes h) \\ &= \mathcal{F}_2 M (t^2 I - M^*M)^{-1} \begin{bmatrix} I_{\mathcal{E}_1} \\ 0 \\ \vdots \\ 0 \end{bmatrix} h \end{aligned}$$

for any $h \in \mathcal{E}_1$. Similarly, one can get the formula for ψ. Using again Theorem 3.3, we can complete the proof. □

Now, we obtain an explicit solution for the Carathéodory-Schur optimization problem (3.11) for the analytic Toeplitz algebra R_n^∞. Denote

$$N := \text{card}\,\{\alpha \in \mathbb{F}_n^+ : |\alpha| \leq m-1\}.$$

THEOREM 3.8. *Let* $q := \sum_{|\alpha|\leq m-1} a_\alpha R_\alpha$*,* $a_\alpha \in \mathbb{C}$*, be a polynomial in* R_n^∞*, and let* $M := [a_{\alpha,\beta}]_{|\alpha|,|\beta|\leq m-1}$ *be the matrix defined by*

$$a_{\alpha,\beta} := \begin{cases} a_{\alpha\backslash\beta}, & \text{if } \alpha \geq \beta \\ 0, & \text{otherwise.} \end{cases}$$

Let $y \in \mathbb{C}^N$ be any vector which attains the norm of M. Then there is a unique $\varphi \in R_n^\infty$ such that its first N Fourier coefficients are a_α, $|\alpha| \leq m-1$, and $\|\varphi\| = d_\infty = \|M\|$. Moreover, $\frac{1}{d_\infty}\varphi$ is inner and φ is uniquely determined by the equation

$$\varphi \mathcal{F} y = \mathcal{F} M y, \tag{3.15}$$

where $\mathcal{F} : \ell^2(\mathbb{F}_n^+) \to F^2(H_n)$ is the Fourier transform defined by

$$\mathcal{F}[(a_\alpha)_{\alpha \in \mathbb{F}_n^+}] = \sum_{\alpha \in \mathbb{F}_n^+} a_\alpha e_\alpha$$

for any $(a_\alpha)_{\alpha \in \mathbb{F}_n^+} \in \ell^2(\mathbb{F}_n^+)$.

Proof. Let $\mathcal{H}' := \mathcal{P}_{m-1}$, $\mathcal{Y} := \mathbb{C}$, and define the operator

$$A := P_{\mathcal{H}'} q : F^2(H_n) \to \mathcal{H}' \subset F^2(H_n).$$

Applying Theorem 3.5 in this setting, we find $M_\varphi \in R_n^\infty$ such that its first N Fourier coefficients are a_α, $|\alpha| \leq m-1$, and $\|M_\varphi\| = d_\infty = \|M\|$. Moreover $\frac{1}{d_\infty} M_\varphi$ is inner and M_φ is uniquely determined by the equation

$$M_\varphi g = A g, \tag{3.16}$$

where g is any vector in $F^2(H_n)$ where A attains its norm. On the other hand, if $y \in \mathbb{C}^N$ is a vector where M attains its norm, then $\mathcal{F} y$ is a vector where A attains its norm. Therefore, relation (3.16) implies (3.15). The proof is complete. □

We mention that, using Corollary 3.4 and Theorem 3.6, one can obtain multivariable commutative versions of Theorem 3.7 and Theorem 3.8 for $W_n^\infty \bar{\otimes} B(\mathcal{E}_1, \mathcal{E}_2)$ and W_n^∞, respectively. On the other hand, Theorem 3.8 can be extended to polynomials $q \in R_n^\infty \bar{\otimes} B(\mathbb{C}, \mathcal{E}_2)$, where $\mathcal{E}_2$ is an arbitrary Hilbert space.

3.3. Maximal entropy solution for the Nevanlinna-Pick interpolation problem with operatorial argument in several variables

In this section, we obtain explicit forms for the maximal entropy solution of the left tangential Nevanlinna-Pick (see [**35**]) interpolation problem with operatorial argument in several variables (see Theorem 3.9) as well as for its entropy (Theorem 3.11).

As in [**55**], the spectral radius of a sequence of operators $\mathcal{Z} := (Z_1, \ldots, Z_n)$, $Z_i \in B(\mathcal{Y})$, is given by

$$r(\mathcal{Z}) := \lim_{k \to \infty} \| \sum_{|\alpha|=k} Z_\alpha Z_\alpha^* \|^{1/2k} = \inf_{k \to \infty} \| \sum_{|\alpha|=k} Z_\alpha Z_\alpha^* \|^{1/2k}.$$

Note that if $Z_1 Z_1^* + \cdots + Z_n Z_n^* < r I_{\mathcal{Y}}$ with $0 < r < 1$, then $r(\mathcal{Z}) < 1$. Any element f in $F_n^\infty \bar{\otimes} B(\mathcal{H}, \mathcal{Y})$ has a unique Fourier representation

$$f \sim \sum_{\alpha \in \mathbb{F}_n^+} S_\alpha \otimes A_{(\alpha)}$$

for some operators $A_{(\alpha)} \in B(\mathcal{H}, \mathcal{Y})$ such that

$$\sum_{\alpha \in \mathbb{F}_n^+} A_{(\alpha)}^* A_{(\alpha)} \leq \|f\|^2 I.$$

If $r(\mathcal{Z}) < 1$, it makes sense to define *the evaluation* of f at $(Z_1, \ldots, Z_n)$ by setting

$$f(Z_1, \ldots, Z_n) := \sum_{k=0}^{\infty} \sum_{|\alpha|=k} Z_{\tilde{\alpha}} A_{(\alpha)}. \tag{3.17}$$

Now, using the fact that the spectral radius of $\mathcal{Z}$ is strictly less than 1, one can prove the norm convergence of the series (3.17).

Given $C \in B(\mathcal{H}, \mathcal{Y})$, we define the controllability operator $W_{\{\mathcal{Z},C\}} : F^2(H_n) \otimes \mathcal{H} \to \mathcal{Y}$ associated with $\{\mathcal{Z}, C\}$ by setting

$$W_{\{\mathcal{Z},C\}} \left(\sum_{\alpha \in \mathbb{F}_n^+} e_\alpha \otimes h_\alpha \right) := \sum_{k=0}^{\infty} \sum_{|\alpha|=k} Z_{\tilde{\alpha}} C h_\alpha,$$

where $\tilde{\alpha}$ is the reverse of $\alpha := g_{i_1} \cdots g_{i_k} \in \mathbb{F}_n^+$, i.e., $\tilde{\alpha} := g_{i_k} \cdots g_{i_1}$. Since $r(\mathcal{Z}) < 1$, note that $W_{\{\mathcal{Z},C\}}$ is a well-defined bounded operator. We call the positive operator $G_{\{\mathcal{Z},C\}} := W_{\{\mathcal{Z},C\}} W_{\{\mathcal{Z},C\}}^*$ the controllability grammian for $\{\mathcal{Z}, C\}$. It is easy to see that

$$G_{\{\mathcal{Z},C\}} = \sum_{k=0}^{\infty} \sum_{|\alpha|=k} Z_\alpha C C^* Z_\alpha^*, \tag{3.18}$$

where the series converges in norm. As in the classical case ($n = 1$), we say that the pair $\{\mathcal{Z}, C\}$ is controllable if its grammian $G_{\{\mathcal{Z},C\}}$ is strictly positive. We remark that $G_{\{\mathcal{Z},C\}}$ is the unique positive solution of the Lyapunov equation

$$X = \sum_{i=1}^{n} Z_i X Z_i^* + CC^*. \tag{3.19}$$

Given $\mathcal{Z} := [Z_1 \ \cdots \ Z_n] \in B(\oplus_{i=1}^n \mathcal{Y}, \mathcal{Y})$ with $r(\mathcal{Z}) < 1$, and the operators $B \in B(\mathcal{K}, \mathcal{Y})$ and $C \in B(\mathcal{H}, \mathcal{Y})$, we note that

$$\begin{aligned} W_{\{\mathcal{Z},B\}}(R_i \otimes I_\mathcal{K}) &= Z_i W_{\{\mathcal{Z},B\}} \quad \text{and} \\ W_{\{\mathcal{Z},C\}}(R_i \otimes I_\mathcal{H}) &= Z_i W_{\{\mathcal{Z},C\}} \quad \text{for any } i = 1, \ldots, n. \end{aligned} \tag{3.20}$$

On the other hand, if $\Phi \in F_n^\infty \bar{\otimes} B(\mathcal{H}, \mathcal{K})$ has the Fourier representation

$$\Phi \sim \sum_{\alpha \in \mathbb{F}_n^+} S_\alpha \otimes A_{(\alpha)},$$

then

$$[(I \otimes B)\Phi](\mathcal{Z}) = C \tag{3.21}$$

if and only if

$$\sum_{k=0}^{\infty} \sum_{|\alpha|=k} Z_{\tilde{\alpha}} B A_{(\alpha)} = C.$$

A straightforward computation on the elements of the form $e_\beta \otimes h$, $h \in \mathcal{H}$, $\beta \in \mathbb{F}_n^+$, shows that relation (3.21) holds if and only if

$$W_{\{\mathcal{Z},B\}} \Phi = W_{\{\mathcal{Z},C\}}. \tag{3.22}$$

Consider the optimization problem

$$d_\infty(\mathcal{Z}, B, C) := \inf \left\{ \|\Theta\| : \ [(I \otimes B)\Theta](\mathcal{Z}) = C, \ \Theta \in F_n^\infty \bar{\otimes} B(\mathcal{H}, \mathcal{K}) \right\}, \tag{3.23}$$

which we call the standard left tangential Nevanlinna-Pick interpolation problem with multivariable operatorial argument. According to Theorem 7.2 from [**55**], we have $d_\infty := d_\infty(\mathcal{Z}, B, C) < \infty$, if and only if there exists a bounded operator

$$A : F^2(H_n) \otimes \mathcal{H} \to \mathcal{H}' := \overline{\text{range } W^*_{\{\mathcal{Z},B\}}} \subseteq F^2(H_n) \otimes \mathcal{K}$$

such that

$$W_{\{\mathcal{Z},B\}}A = W_{\{\mathcal{Z},C\}}. \tag{3.24}$$

Moreover, in this case A is uniquely determined and there exists an operator $\Theta_{\text{opt}} \in F_n^\infty \bar{\otimes} B(\mathcal{H}, \mathcal{K})$ such that $[(I \otimes B)\Theta_{\text{opt}}](\mathcal{Z}) = C$ and

$$d_\infty = \|A\| = \|\Theta_{\text{opt}}\|. \tag{3.25}$$

Setting $T_i' := P_{\mathcal{H}'}(R_i \otimes I_\mathcal{K})|\mathcal{H}'$, we have $T_i'A = A(R_i \otimes I_\mathcal{H})$ for any $i = 1, \ldots, n$. It is easy to see that $\hat{A}$ is an intertwining lifting of A with respect to $\{R_i \otimes I_\mathcal{K}\}_{i=1}^n$ if and only if $\hat{A} = \Theta$ for some $\Theta \in F_n^\infty \bar{\otimes} B(\mathcal{H}, \mathcal{K})$ such that $W_{\{\mathcal{Z},B\}}\Theta = W_{\{\mathcal{Z},C\}}$. Using this fact, one can see that the optimization problem (3.23) is equivalent to the following

$$d_\infty(\mathcal{Z}, B, C) := \inf\left\{\|\hat{A}\| : \ \hat{A} \text{ is an intertwining lifting of } A\right\}. \tag{3.26}$$

On the other hand, we say that $\Theta_c \in F_n^\infty \bar{\otimes} B(\mathcal{H}, \mathcal{K})$ is the central interpolant for the standard Nevanlinna-Pick problem with tolerance $t > 0$ if Θ_c is the central intertwining lifting of A with tolerance t, i.e., $\|\Theta_c\| \leq t$. In [**55**], we proved that, given $t > 0$, there exists $\Theta \in F_n^\infty \bar{\otimes} B(\mathcal{H}, \mathcal{K})$ satisfying

$$[(I \otimes B)\Theta](\mathcal{Z}) = C \quad \text{and} \quad \|\Theta\| \leq t, \tag{3.27}$$

if and only if

$$t^2 G_{\{\mathcal{Z},B\}} - G_{\{\mathcal{Z},C\}} \geq 0,$$

where $G_{\{\mathcal{Z},B\}}$ and $G_{\{\mathcal{Z},C\}}$ are the grammians for $\{\mathcal{Z}, B\}$ and $\{\mathcal{Z}, C\}$, respectively. Now, using the results from previous sections, we deduce that Θ_c is the maximal entropy of the interpolation problem (3.27), if $\dim \mathcal{H} < \infty$.

In what follows, assume that the grammian $G_{\{\mathcal{Z},B\}}$ is strictly positive. According to [**55**], there exists $\Theta_{\text{opt}} \in F_n^\infty \bar{\otimes} B(\mathcal{H}, \mathcal{K})$ solving the problem (3.23), i.e.,

$$[(I \otimes B)\Theta_{\text{opt}}](\mathcal{Z}) = C \quad \text{and} \quad d_\infty = \|\Theta_{\text{opt}}\|. \tag{3.28}$$

Moreover, the spectral radius of $G_{\{\mathcal{Z},C\}}G^{-1}_{\{\mathcal{Z},B\}}$ is equal to d_∞^2.

Now, we can obtain an explicit form for the maximal entropy solution of the standard left tangential Nevanlinna-Pick interpolation problem with operatorial argument in several variables.

THEOREM 3.9. *Let* $\mathcal{Z} := [Z_1 \ \cdots \ Z_n] \in B(\oplus_{i=1}^n \mathcal{Y}, \mathcal{Y})$, $B \in B(\mathcal{K}, \mathcal{Y})$, *and* $C \in B(\mathcal{H}, \mathcal{Y})$ *be operators such that* $r(\mathcal{Z}) < 1$ *and the grammian* $G_{\{\mathcal{Z},B\}}$ *is strictly positive. Then* $d_\infty < \infty$ *and the central interpolant* Θ_t *with tolerance* $t > d_\infty$ *for the standard left tangential Nevanlinna-Pick problem with operatorial argument is given by*

$$\Theta_t = \Phi\Psi^{-1},$$

where $\Phi \in F_n^\infty \bar{\otimes} B(\mathcal{H},\mathcal{K})$ *and* $\Psi \in F_n^\infty \bar{\otimes} B(\mathcal{H})$ *are analytic Toeplitz operators defined by*

$$\Phi := \sum_{\alpha\in\mathbb{F}_n^+} S_\alpha \otimes B^*(Z_{\tilde{\alpha}})^*(t^2 G_{\{\mathcal{Z},B\}} - G_{\{\mathcal{Z},C\}})^{-1} C \quad \text{and}$$

$$\Psi := \frac{1}{t^2}\left[I + \sum_{\alpha\in\mathbb{F}_n^+} S_\alpha \otimes C^*(Z_{\tilde{\alpha}})^*(t^2 G_{\{\mathcal{Z},B\}} - G_{\{\mathcal{Z},C\}})^{-1} C.\right]$$

In particular, $\Theta_t \in F_n^\infty \bar{\otimes} B(\mathcal{H},\mathcal{K})$ *is an analytic Toeplitz operator such that*

$$[(I\otimes B)\Theta_t](Z_1,\dots,Z_n) = C \quad \text{and} \quad \|\Theta_t\| < t.$$

PROOF. Since the grammian $G_{\{\mathcal{Z},B\}}$ is strictly positive, the operator $W^*_{\{\mathcal{Z},B\}}$ has closed range and

$$P_{\text{range } W^*_{\{\mathcal{Z},B\}}} = W^*_{\{\mathcal{Z},B\}} G^{-1}_{\{\mathcal{Z},B\}} W_{\{\mathcal{Z},B\}}.$$

Note that the operator

$$A := W^*_{\{\mathcal{Z},B\}} G^{-1}_{\{\mathcal{Z},B\}} W_{\{\mathcal{Z},C\}} \tag{3.29}$$

satisfies the equation $W_{\{\mathcal{Z},B\}} A = W_{\{\mathcal{Z},C\}}$ and

$$A(R_i\otimes I_{\mathcal{H}}) = AT_i, \quad i=1,\dots,n,$$

where

$$T_i' := P_{\mathcal{H}'}(R_i\otimes I_{\mathcal{K}})|\mathcal{H}' \quad \text{and} \quad \mathcal{H}' := \text{range } W^*_{\{\mathcal{Z},B\}}.$$

Since $d_\infty = \|A\|$, we have $\|A\| < t$. On the other hand, since

$$W_{\{\mathcal{Z},B\}} AA^* W^*_{\{\mathcal{Z},B\}} = W_{\{\mathcal{Z},C\}} W^*_{\{\mathcal{Z},C\}},$$

it is easy to see that the operator $t^2 G_{\{\mathcal{Z},B\}} - G_{\{\mathcal{Z},C\}}$ is strictly positive. Let B_c be the central intertwining lifting of A with respect to $\{R_i\otimes I_{\mathcal{K}}\}_{i=1}^n$ and tolerance t. Then $B_c \in F_n^\infty \bar{\otimes} B(\mathcal{H},\mathcal{K})$, $\|B_c\| \le t$, and $W_{\{\mathcal{Z},B\}} B_c = W_{\{\mathcal{Z},C\}}$.

Now, we use Theorem 3.3 to find an explicit formula for B_c. To this end, according to relation (3.29), straightforward calculations show that

$$D^2_{A,t} := t^2 I - A^*A = t^2 I - W^*_{\{\mathcal{Z},C\}} G^{-1}_{\{\mathcal{Z},B\}} W_{\{\mathcal{Z},C\}}$$

and

$$D^{-2}_{A,t} = \frac{1}{t^2}\left[I + W^*_{\{\mathcal{Z},C\}}\left(t^2 G_{\{\mathcal{Z},B\}} - G_{\{\mathcal{Z},C\}}\right)^{-1} W_{\{\mathcal{Z},C\}}\right]. \tag{3.30}$$

Hence, and using again relation (3.29), we obtain

$$AD^{-2}_{A,t} = W^*_{\{\mathcal{Z},B\}}\left(t^2 G_{\{\mathcal{Z},B\}} - W_{\{\mathcal{Z},C\}}\right)^{-1} W_{\{\mathcal{Z},C\}}.$$

Note that $W_{\{\mathcal{Z},C\}}(1\otimes h) = Ch$, $h\in\mathcal{H}$, and

$$W^*_{\{\mathcal{Z},C\}} y = \sum_{\alpha\in\mathbb{F}_n^+} e_\alpha \otimes C^*(Z_{\tilde{\alpha}})^* y, \quad y\in\mathcal{Y}.$$

Therefore, the above equations imply

$$D^{-2}_{A,t}(1\otimes h) = \frac{1}{t^2}\left[1\otimes h + \sum_{\alpha\in\mathbb{F}_n^+} e_\alpha \otimes C^*(Z_{\tilde{\alpha}})^*\left(t^2 G_{\{\mathcal{Z},B\}} - G_{\{\mathcal{Z},C\}}\right)^{-1} Ch\right]$$

and

$$AD_{A,t}^{-2}(1\otimes h)=\sum_{\alpha\in\mathbb{F}_n^+} e_\alpha\otimes B^*(Z_{\tilde\alpha})^*\left(t^2G_{\{\mathcal{Z},B\}}-G_{\{\mathcal{Z},C\}}\right)^{-1}Ch$$

for any $h\in\mathcal{H}$.

Now applying Theorem 3.3 to the operator A, which satisfies $\|A\|<t$, we find the explicit forms for Φ and Ψ mentioned in the theorem. According to Corollary 2.16 and Theorem 2.18, the operator Ψ is invertible in $F_n^\infty\bar\otimes B(\mathcal{H})$ and

$$t^2I-\Theta_t^*\Theta_t=\Psi^*\Psi.$$

This implies $\|\Theta_t\|<t$, and the proof is complete. □

REMARK 3.10. *Under the conditions of Theorem 3.9, the operators Φ and Ψ satisfy the following state space formulas*

$$\Phi=(I\otimes B^*)\left(I-\sum_{i=1}^n S_i\otimes Z_i^*\right)^{-1}\left(I\otimes(t^2G_{\{\mathcal{Z},B\}}-G_{\{\mathcal{Z},C\}})^{-1}C\right),$$

$$\Psi=\frac{1}{t^2}\left[I+(I\otimes C^*)\left(I-\sum_{i=1}^n S_i\otimes Z_i^*\right)^{-1}\left(I\otimes(t^2G_{\{\mathcal{Z},B\}}-G_{\{\mathcal{Z},C\}})^{-1}C\right)\right].$$

PROOF. Note that, for any $k=1,2,\ldots$, the operators S_α, $|\alpha|=k$, are isometries with orthogonal ranges. Now, it is easy to see that

$$\begin{aligned}\|(S_1\otimes Z_1^*+\cdots+S_n\otimes Z_n^*)^k\| &= \left\|\sum_{|\alpha|=k} S_\alpha\otimes Z_{\tilde\alpha}^*\right\|\\ &=\left\|\left(\sum_{|\alpha|=k}S_\alpha^*\otimes Z_{\tilde\alpha}\right)\left(\sum_{|\alpha|=k}S_\alpha\otimes Z_{\tilde\alpha}^*\right)\right\|^{1/2}\\ &=\left\|\sum_{|\alpha|=k} I\otimes Z_{\tilde\alpha}Z_{\tilde\alpha}^*\right\|^{1/2}=\left\|\sum_{|\alpha|=k}Z_\alpha Z_\alpha^*\right\|^{1/2}.\end{aligned}$$

Since $\mathrm{r}(\mathcal{Z})<1$, the root test implies the convergence of the series

$$\sum_{k=0}^\infty\|(S_1\otimes Z_1^*+\cdots+S_n\otimes Z_n^*)^k\|.$$

Hence, it is clear that

$$(I-S_1\otimes Z_1^*-\ldots-S_n\otimes Z_n^*)^{-1}=\sum_{k=0}^\infty\sum_{|\alpha|=k}S_\alpha\otimes Z_{\tilde\alpha}^*.$$

Using Theorem 3.9, we can complete the proof. □

Let $\Theta\in F_n^\infty\bar\otimes B(\mathcal{H},\mathcal{K})$ be an interpolant for the standard Nevanlinna-Pick problem with tolerance t, i.e.,

$$[(I\otimes B)\Theta](\mathcal{Z})=C\quad\text{and}\quad\|\Theta\|\le t. \tag{3.31}$$

Define

$$\langle\Delta(\Theta)x,x\rangle=\inf\{\langle(t^2I-\Theta^*\Theta)(x-p),x-p\rangle:\ p\in F^2(H_n)\otimes\mathcal{H}\ \text{and}\ p(0)=0\}$$

for any $x \in \mathcal{H}$. If $\mathcal{H}$ is finite dimensional, then the entropy of Θ is defined by

$$E(\Theta) := \ln \det \Delta(\Theta).$$

According to Theorem 2.9, we have $\Delta(\Psi) \leq \Delta(\Theta_t)$, where Θ_t is the central interpolant for the problem (3.31) and Ψ is any interpolant for the same problem. Moreover, if $\dim \mathcal{H} < \infty$, then Corollary 2.21 shows that $E(\Theta_t) \geq E(\Psi)$, i.e., Θ_t is a maximal entropy interpolant.

Now, we can prove the following result.

THEOREM 3.11. *Let $\mathcal{Z} := [Z_1 \ \cdots \ Z_n] \in B(\oplus_{i=1}^n \mathcal{Y}, \mathcal{Y})$, $B \in B(\mathcal{K}, \mathcal{Y})$, and $C \in B(\mathcal{H}, \mathcal{Y})$ be operators such that $r(\mathcal{Z}) < 1$ and the grammian $G_{\{\mathcal{Z},B\}}$ is strictly positive. Then the central interpolant Θ_t with tolerance $t > d_\infty(\mathcal{Z}, B, C)$ is the maximal interpolant for the standard left Nevanlinna-Pick problem with operatorial argument and tolerance t, and*

$$\Delta(\Theta_t) = \frac{1}{t^2}\left[I_{\mathcal{H}} + C^*\left(t^2 G_{\{\mathcal{Z},B\}} - G_{\{\mathcal{Z},C\}}\right)^{-1} C\right]. \tag{3.32}$$

If $\dim \mathcal{H} < \infty$, then the entropy of Θ_t is given by

$$E(\Theta_t) = -\ln \det \left\{\frac{1}{t^2}\left[I_{\mathcal{H}} + C^*\left(t^2 G_{\{\mathcal{Z},B\}} - G_{\{\mathcal{Z},C\}}\right)^{-1} C\right]\right\} \tag{3.33}$$

Moreover, Θ_t is the maximal entropy interpolant, i.e., $E(\Psi) \leq E(\Theta_t)$ for any interpolant Ψ, and $E(\Theta_t) = E(\Psi)$ if and only if $\Theta_t = \Psi$.

PROOF. According to Theorem 2.9 (when $\|A\| < t$) and Theorem 3.9, we have

$$\Delta(\Theta_t) = \Delta(A) = P_{\mathcal{H}} D_{A,t}^{-1} | \mathcal{H}.$$

Using relation (3.30) and the definition of $W_{\{\mathcal{Z},C\}}$ and $W_{\{\mathcal{Z},B\}}$, we get equation (3.32). Now, we can use Theorem 2.18 to obtain (3.33). The last part of the theorem follows by applying Corollary 2.21 to our setting. The proof is complete. □

Let $\mathcal{H}$, $\mathcal{K}$, and $\mathcal{Y}_i$, $i = 1, \ldots, m$, be Hilbert spaces and consider the operators

$$\begin{aligned} &B_j : \mathcal{K} \to \mathcal{Y}_j, \quad C_j : \mathcal{H} \to \mathcal{Y}_j, \quad j = 1, \ldots, m, \\ &\Lambda_j := [\Lambda_{j1} \ \cdots \ \Lambda_{jn}] : \oplus_{i=1}^n \mathcal{Y}_j \to \mathcal{Y}_j, \quad j = 1, \ldots, m, \end{aligned} \tag{3.34}$$

such that $r(\Lambda_j) < 1$ for any $j = 1, \ldots, m$. The left tangential Nevanlinna-Pick interpolation problem with operatorial argument and tolerance $t > 0$ for the tensor product $F_n^\infty \bar{\otimes} B(\mathcal{H}, \mathcal{K})$ is to find Φ in $F_n^\infty \bar{\otimes} B(\mathcal{H}, \mathcal{K})$ such that $\|\Phi\| \leq t$ and

$$[(I \otimes B_j)\Phi](\Lambda_j) = C_j, \quad j = 1, \ldots, m. \tag{3.35}$$

Define the following operators

$$B := \begin{bmatrix} B_1 \\ \vdots \\ B_m \end{bmatrix} : \mathcal{K} \to \oplus_{j=1}^m \mathcal{Y}_j, \quad C := \begin{bmatrix} C_1 \\ \vdots \\ C_m \end{bmatrix} : \mathcal{H} \to \oplus_{j=1}^m \mathcal{Y}_j,$$

and $\mathcal{Z} := [Z_1 \ \cdots \ Z_n]$, where Z_i is the diagonal operator defined by

$$Z_i := \begin{bmatrix} \Lambda_{1i} & 0 & 0 \\ 0 & \Lambda_{2i} & 0 \\ \vdots & \vdots & \vdots \\ 0 & 0 & \Lambda_{mi} \end{bmatrix} : \oplus_{j=1}^m \mathcal{Y}_j \to \oplus_{j=1}^m \mathcal{Y}_j$$

for any $i = 1, \ldots, n$. Note that the interpolation relation (3.35) is equivalent to relation (3.21) and we have

$$t^2 G_{\{\mathcal{Z},B\}} - G_{\{\mathcal{Z},C\}} = \sum_{p=0}^{\infty} \sum_{|\alpha|=p} Z_\alpha [t^2 BB^* - CC^*] Z_\alpha^*$$

$$= \left[\sum_{p=0}^{\infty} \sum_{|\alpha|=p} \Lambda_{j\alpha} [t^2 B_j B_k^* - C_j C_k^*] \Lambda_{k\alpha}^* \right]_{j,k=1}^m$$

where $W_{\{\mathcal{Z},B\}}$ and $W_{\{\mathcal{Z},C\}}$ are the controllability operators associated with $\{\mathcal{Z}, B\}$ and $\{\mathcal{Z}, C\}$, respectively.

It [**55**], we proved that the left tangential Nevanlinna-Pick interpolation problem with data Λ_j, B_j, and C_j, $j = 1, \ldots, m$, and tolerance $t > 0$, has a solution if and only if the operator matrix

$$\left[\sum_{p=0}^{\infty} \sum_{|\alpha|=p} \Lambda_{j\alpha} [t^2 B_j B_k^* - C_j C_k^*] \Lambda_{k\alpha}^* \right]_{j,k=1}^m \tag{3.36}$$

is positive semidefinite.

REMARK 3.12. *All the results of this section can be written for the left tangential Nevanlinna-Pick interpolation problem with operatorial argument with data Λ_j, B_j, and C_j, $j = 1, \ldots, m$, and tolerance $t > 0$. In particular, one can obtain an explicit form of the maximal entropy solution of the above mentioned interpolation problem.*

3.4. Maximal entropy interpolation on the unit ball of $\mathbb{C}^n$

In this section, we present some consequences of the results of this paper to analytic interpolation on the open unit ball $\mathbb{B}_n$ of $\mathbb{C}^n$. Let $z_j := (z_{j1}, \ldots, z_{jn})$, $j = 1, \ldots, m$, be distinct points in $\mathbb{B}_n$, and let $B_j \in B(\mathcal{K}, \mathcal{Y}_j)$, $C_j \in B(\mathcal{H}, \mathcal{Y}_j)$, $j = 1, \ldots, m$. The left tangential Nevanlinna-Pick interpolation problem with data $z_j \in \mathbb{B}_n$, B_j, C_j, $j = 1, \ldots, m$, and tolerance $t > 0$, is to find $\Theta \in F_n^\infty \bar{\otimes} B(\mathcal{H}, \mathcal{K})$ such that

$$B_j \Theta(z_j) = C_j, \quad j = 1, \ldots, m, \tag{3.37}$$

and $\|\Theta\| \leq t$. We proved in [**46**] that this interpolation problem has solution if and only if the operator matrix

$$\left[\frac{t^2 B_j B_k^* - C_j C_k^*}{1 - \langle z_j, z_k \rangle} \right]_{j,k=1}^m$$

is positive semidefinite. Now, as a consequence of the results of Section 3.3, we can find the maximal entropy solution of the interpolation problem (3.37). Moreover, under certain natural conditions, we obtain an explicit form for the unique solution of the Nevanlinna-Pick optimization problem (see Theorem 3.14). Finally, we apply our permanence principle to the Nevanlinna-Pick interpolation problem on the unit ball (see Theorem 3.16).

First, we obtain an explicit form for the maximal entropy solution of the left tangential Nevanlinna-Pick interpolation problem with tolerance t on the unit ball, under certain natural conditions.

THEOREM 3.13. *Let $z_j := (z_{j1}, \ldots, z_{jn})$, $j = 1, \ldots, m$, be distinct points in $\mathbb{B}_n$, and let $B_j \in B(\mathcal{K}, \mathcal{Y}_j)$, $C_j \in B(\mathcal{H}, \mathcal{Y}_j)$, $j = 1, \ldots, m$, be operators such that the grammian*

$$\left[\frac{B_j B_k^*}{1 - \langle z_j, z_k \rangle}\right]_{j,k=1}^m \tag{3.38}$$

is strictly positive. Let $t > d_\infty$ and let Θ_t be the central interpolant for the left tangential Nevanlinna-Pick interpolation problem with tolerance t. Then $\Theta_t(\lambda) = \Phi(\lambda)\Psi(\lambda)^{-1}$ where, for any $\lambda = (\lambda_1, \ldots, \lambda_n) \in \mathbb{B}_n$,

$$\Phi(\lambda_1, \ldots, \lambda_n) = B^* \left(I - \sum_{i=1}^n \lambda_i Z_i^*\right)^{-1} \left(\left[\frac{t^2 B_j B_k^* - C_j C_k^*}{1 - \langle z_j, z_k \rangle}\right]_{j,k=1}^m\right)^{-1} C$$

and

$$\Psi(\lambda_1, \ldots, \lambda_n) = \frac{1}{t^2} \left\{ I + C^* \left(I - \sum_{i=1}^n \lambda_i Z_i^*\right)^{-1} \left(\left[\frac{t^2 B_j B_k^* - C_j C_k^*}{1 - \langle z_j, z_k \rangle}\right]_{j,k=1}^m\right)^{-1} C \right\},$$

where

$$Z_i := \begin{bmatrix} z_{1i} I_{\mathcal{Y}_i} & 0 & \cdots & 0 \\ 0 & z_{2i} I_{\mathcal{Y}_i} & \cdots & 0 \\ 0 & 0 & \cdots & z_{mi} I_{\mathcal{Y}_i} \end{bmatrix}, \quad B = \begin{bmatrix} B_1 \\ \vdots \\ B_m \end{bmatrix}, \quad C = \begin{bmatrix} C_1 \\ \vdots \\ C_m \end{bmatrix} \tag{3.39}$$

for any $i = 1, \ldots, n$. In particular, if $\dim \mathcal{H} < \infty$, then Θ_t is the maximal entropy solution satisfying $\|\Theta_t\| < t$ and $B_j \Theta_t(z_j) = C_j$ for any $j = 1, \ldots, m$. Moreover, the entropy of Θ_t is given by

$$E(\Theta_t) = -\ln \det \left\{ \frac{1}{t^2} \left[I + C^* \left(\left[\frac{t^2 B_j B_k^* - C_j C_k^*}{1 - \langle z_j, z_k \rangle}\right]_{j,k=1}^m\right)^{-1} C \right] \right\}.$$

PROOF. Let $z_j := (z_{j1}, \ldots, z_{jn})$, $j = 1, \ldots, m$, be distinct points in $\mathbb{B}_n$. Note that for any $j, k = 1, \ldots, m$, we have

$$\sum_{\alpha \in \mathbb{F}_n^+} z_{j\alpha} \bar{z}_{k\alpha} = \frac{1}{1 - \langle z_j, z_k \rangle}.$$

A simple computation shows that

$$G_{\{\mathcal{Z}, B\}} = \left[\sum_{\alpha \in \mathbb{F}_n^+} z_{j\alpha} \bar{z}_{k\alpha} B_j B_k^*\right]_{j,k=1}^m = \left[\frac{B_j B_k^*}{1 - \langle z_j, z_k \rangle}\right]_{j,k=1}^m.$$

Therefore, we have

$$t^2 G_{\{\mathcal{Z}, B\}} - G_{\{\mathcal{Z}, C\}} = \left[\frac{t^2 B_j B_k^* - C_j C_k^*}{1 - \langle z_j, z_k \rangle}\right]_{j,k=1}^m,$$

where $\mathcal{Z} := [Z_1 \ \cdots \ Z_n]$. Since $\mathrm{r}(\mathcal{Z}) < 1$ and the grammian $G_{\{\mathcal{Z}, B\}}$ is strictly positive, we have $d_\infty < \infty$ (see Section 3.3). Now, we can apply Theorem 3.9 and find the central interpolant Θ_t for the left tangential Nevanlinna-Pick interpolation problem with tolerance $t > d_\infty$, on the unit ball. Using Remark 3.10, and taking the compression of Θ_t to the symmetric Fock space, we get the corresponding formulas for $\Phi(\lambda_1,\dots,\lambda_n)$ and $\Psi(\lambda_1,\dots,\lambda_n)$. To complete the proof of the theorem, we need now to use Theorem 3.11 in our particular setting. □

As mentioned in Section 3.3, if the grammian $G_{\{\mathcal{Z},B\}}$ is strictly positive, then there exists $\Theta_{\text{opt}} \in F_n^\infty \bar{\otimes} B(\mathcal{H},\mathcal{K})$ solving the problem

$$[(I\otimes B)\Theta_{\text{opt}}](\mathcal{Z}) = C \quad \text{and} \quad d_\infty = \|\Theta_{\text{opt}}\|. \tag{3.40}$$

In what follows, we find an explicit form for the unique solution of the optimization problem (3.40) on the unit ball $\mathbb{B}_n$.

THEOREM 3.14. *Let $z_j := (z_{j1},\dots,z_{jn})$, $j=1,\dots,m$, be distinct points in $\mathbb{B}_n$, and Z_i be defined as in (3.39). Let $B_j \in B(\mathcal{K},\mathcal{Y}_j)$ and $C_j\in B(\mathbb{C},\mathcal{Y}_j)$, $j=1,\dots,m$, be operators with $\dim \mathcal{Y}_j<\infty$ and such that the grammian*

$$\left[\frac{B_jB_k^*}{1-\langle z_j,z_k\rangle}\right]_{j,k=1}^m$$

is strictly positive. Then the unique solution $\Theta_{\text{opt}} \in F_n^\infty\bar{\otimes} B(\mathbb{C},\mathcal{K})$ for the Nevanlinna-Pick optimization problem satisfies the equation

$$\Theta_{\text{opt}}(\lambda_1,\dots,\lambda_n) = \frac{d_\infty^2[B_1^* \ \cdots \ B_n^*](I-\lambda_1Z_1^*-\cdots-\lambda_nZ_n^*)^{-1}x}{[C_1^* \ \cdots C_n^*](I-\lambda_1Z_1^*-\cdots-\lambda_nZ_n^*)^{-1}x}, \tag{3.41}$$

where $\lambda=(\lambda_1,\dots,\lambda_n)\in\mathbb{B}_n$, Z_i is given by (3.39), and x is the eigenvector corresponding to the largest eigenvalue $\lambda_{\max}$ of the operator

$$\left(\left[\frac{B_jB_k^*}{1-\langle z_j,z_k\rangle}\right]_{j,k=1}^m\right)^{-1}\left[\frac{C_jC_k^*}{1-\langle z_j,z_k\rangle}\right]_{j,k=1}^m.$$

Moreover, $d_\infty^2=\lambda_{\max}$ and $\frac{1}{d_\infty}\Theta_{\text{opt}}$ is inner in $F_n^\infty\bar{\otimes}B(\mathbb{C},\mathcal{K})$. In particular,

$$\|\Theta_{\text{opt}}\| = d_\infty, \quad B_j\Theta_{\text{opt}}(z_j)=C_j, \quad \textit{for any } j=1,\dots,m.$$

PROOF. Let $\mathcal{Z} := [Z_1 \ \cdots \ Z_n]\in B(\oplus_{i=1}^n\mathcal{Y},\mathcal{Y})$, $B\in B(\mathcal{K},\mathcal{Y})$, and $C\in B(\mathbb{C},\mathcal{Y})$ be the data for the standard left Nevanlinna-Pick interpolation problem, where $\dim\mathcal{Y}<\infty$. First we prove that if $\mathrm{r}(\mathcal{Z})<1$ and $G_{\{\mathcal{Z},B\}}$ is strictly positive, then there is a unique solution $\Theta\in F_n^\infty\bar{\otimes}B(\mathbb{C},\mathcal{K})$ for the interpolation problem (3.28), given by the equation

$$\Theta\left(\sum_{\alpha\in\mathbb{F}_n^+} e_\alpha\otimes C^*Z_{\tilde{\alpha}}^*x\right) = d_\infty^2\sum_{\alpha\in\mathbb{F}_n^+} e_\alpha\otimes B^*Z_{\tilde{\alpha}}^*x, \tag{3.42}$$

where x is the eigenvector corresponding to the largest eigenvalue $\lambda_{\max}$ of the operator $G_{\{\mathcal{Z},B\}}^{-1}G_{\{\mathcal{Z},C\}}\in B(\mathcal{Y})$. Moreover, $d_\infty^2=\lambda_{\max}$ and $\frac{1}{d_\infty}\Theta$ is inner.

Indeed, we know that if $G_{\{\mathcal{Z},B\}}$ is strictly positive, then the operator

$$A = W_{\{\mathcal{Z},B\}}^*G_{\{\mathcal{Z},B\}}^{-1}W_{\{\mathcal{Z},C\}}$$

satisfies relation (3.24). Let x be an eigenvector corresponding to the largest eigenvalue $\lambda_{\max}$ of $G_{\{\mathcal{Z},B\}}^{-1}G_{\{\mathcal{Z},C\}}$. Using the above form of A, we get

$$\begin{aligned} A^*AW_{\{\mathcal{Z},C\}}^* &= A^*A(A^*W_{\{\mathcal{Z},B\}}^*x) = A^*W_{\{\mathcal{Z},B\}}^*G_{\{\mathcal{Z},B\}}^{-1}G_{\{\mathcal{Z},C\}}x \\ &= \lambda_{\max}A^*W_{\{\mathcal{Z},B\}}^*x = \lambda_{\max}W_{\{\mathcal{Z},C\}}^*x. \end{aligned}$$

On the other hand, note that $W^*_{\{\mathcal{Z},C\}}x \neq 0$ because $W^*_{\{\mathcal{Z},B\}}x \neq 0$ is the eigenvector corresponding to the largest eigenvalue $\lambda_{\max}$ of AA^*. Since $\mathcal{Y}$ is finite dimensional, the operator A attains its norm at the vector $g := W^*_{\{\mathcal{Z},C\}}x$.

Now, we use Theorem 3.5 to solve the optimization problem (3.26). First, note that the operator $G^*_{\{\mathcal{Z},B\}}$ is one-to-one and onto $\mathcal{H}'$ (see Section 3.3 for the definition of $\mathcal{H}'$). Now, one can easily see that AA^* is similar to $G^{-1}_{\{\mathcal{Z},B\}}G_{\{\mathcal{Z},C\}}$. In particular, if $\lambda_{\max}$ is the largest eigenvalue of $G^{-1}_{\{\mathcal{Z},B\}}G_{\{\mathcal{Z},C\}}$, then

$$\lambda_{\max} = \|AA^*\| = \|A\|^2 = d_\infty^2.$$

Note that

$$W^*_{\{\mathcal{Z},C\}}x = \sum_{\alpha\in\mathbb{F}_n^+} e_\alpha \otimes C^* Z^*_{\tilde{\alpha}}x$$

and

$$\begin{aligned} AW^*_{\{\mathcal{Z},C\}}x &= W^*_{\{\mathcal{Z},B\}}G^{-1}_{\{\mathcal{Z},B\}}G_{\{\mathcal{Z},C\}}x = \lambda_{\max}W^*_{\{\mathcal{Z},B\}}x \\ &= \lambda_{\max}\sum_{\alpha\in\mathbb{F}_n^+} e_\alpha \otimes B^* Z^*_{\tilde{\alpha}}x. \end{aligned}$$

Therefore, according to the R_n^∞-version of Theorem 3.5, the equation (3.42) holds. Since $\mathrm{r}(\mathcal{Z}) < 1$, as in the proof of Remark 3.10, the equation (3.42) becomes

$$\begin{aligned} &\Theta[(I\otimes C^*)(I - S_1\otimes Z_1^* - \cdots - S_n\otimes Z_n^*)^{-1}(1\otimes x)) \\ &\qquad = d_\infty(I\otimes B^*)(I - S_1\otimes Z_1^* - \cdots - S_n\otimes Z_n^*)^{-1}(1\otimes x). \end{aligned}$$

Since the symmetric Fock space $F_s^2(H_n)$ is invariant under each operator $S_1^*,\ldots,S_n^*$, we can multiply the previous equation to the left by the orthogonal projection $P_{F_s^2(H_n)\otimes\mathcal{K}}$ and obtain

$$\begin{aligned} &\Theta(\lambda_1,\ldots,\lambda_n)C^*(I - \lambda_1 Z_1^* - \cdots - \lambda_n Z_n^*)^{-1}x \\ &\qquad = d_\infty^2 B^*(I - \lambda_1 Z_1^* - \cdots - \lambda_n Z_n^*)^{-1}x \end{aligned}$$

for any $(\lambda_1,\ldots,\lambda_n)\in\mathbb{B}_n$. Here, we used the identification of W_n^∞ with the algebra of analytic multipliers of the symmetric Fock space $F_s^2(H_n)$ (see [**6**]). Therefore, setting $\Theta_{\mathrm{opt}} := \Theta$, the relation (3.41) follows. The proof is complete. □

We present now an application of the permanence principle to the Nevanlinna-Pick interpolation problem on the unit ball $\mathbb{B}_n$. Let $z_1,\ldots,z_k$ be distinct points in the unit ball $\mathbb{B}_n$, and let $C_j\in B(\mathcal{K},\mathcal{K}')$, $j=1,\ldots,k$. The Nevanlinna-Pick interpolation problem with tolerance t for $R_n^\infty\bar{\otimes}B(\mathcal{K},\mathcal{K}')$ is to find $\Phi\in R_n^\infty\bar{\otimes}B(\mathcal{K},\mathcal{K}')$ such that

$$\|\Phi\|\le t \ \text{ and } \ \Phi(z_j) = C_j,\ j=1,\ldots,k. \tag{3.43}$$

According to [**53**], this problem has solutions if and only $\|P_{\mathcal{N}'}\Psi\|\le t$, where $\Psi\in R_n^\infty\bar{\otimes}B(\mathcal{K},\mathcal{K}')$ is an arbitrary operator such that $\Psi(z_j) = C_j$, $j=1,\ldots,k$, and

$$\mathcal{N}' := \operatorname{span}\{f_{z_j}:\ j=1,\ldots,k\}\otimes\mathcal{K}', \tag{3.44}$$

where

$$f_\lambda := (I - \bar{\lambda}_1 S_1 - \cdots - \bar{\lambda}_n S_n)^{-1}(1)\in F^2(H_n)$$

for any $\lambda = (\lambda_1,\ldots,\lambda_n)\in\mathbb{B}_n$.

In what follows we need the following factorization result for contractive multi-analytic operators.

LEMMA 3.15. *Let $\Theta \in R_n^\infty \bar{\otimes} B(\mathcal{E}, \mathcal{E}')$ be a contractive multi-analytic operator. Then Θ admits a unique decomposition $\Theta = \Phi \oplus \Lambda$ with the following properties:*

(i) *$\Psi \in R_n^\infty \bar{\otimes} B(\mathcal{E}_0, \mathcal{E}_0')$ is purely contractive, i.e., $\|P_{\mathcal{E}_0'} \Psi h\| < \|h\|$ for any $h \in \mathcal{E}_0$, $h \neq 0$;*
(ii) *$\Lambda = I \otimes U \in R_n^\infty \bar{\otimes} B(\mathcal{E}_u, \mathcal{E}_u')$, where $U \in B(\mathcal{E}_u, \mathcal{E}_u')$ is a unitary operator;*
(iii) *$\mathcal{E} = \mathcal{E}_0 \oplus \mathcal{E}_u$ and $\mathcal{E}' = \mathcal{E}_0' \oplus \mathcal{E}_u'$.*

PROOF. Let $\Theta = \sum_{\alpha \in \mathbb{F}_n^+} R_\alpha \otimes \theta_\alpha$, $\theta_\alpha \in B(\mathcal{E}, \mathcal{E}')$, be the Fourier representation of Θ. It is well-known that any contraction $\theta_{g_0} \in B(\mathcal{E}, \mathcal{E}')$ admits a unique decomposition $\theta_{g_0} = Z_0 \oplus Z_u$, where $Z_0 \in B(\mathcal{E}_0, \mathcal{E}_0')$ is a pure contraction, i.e., $\|Z_0 h\| < \|h\|$ for any $h \in \mathcal{E}_0$, $h \neq 0$, the operator $Z_u \in B(\mathcal{E}_u, \mathcal{E}_u')$ is unitary, and we have the orthogonal decompositions $\mathcal{E} = \mathcal{E}_0 \oplus \mathcal{E}_u$ and $\mathcal{E}' = \mathcal{E}_0' \oplus \mathcal{E}_u'$. Since Z_u is unitary and Θ is contractice, we deduce

$$\begin{aligned} \|h\|^2 = \|Z_u h\|^2 &= \|\theta_{g_0} h\|^2 \\ &\leq \sum_{\alpha \in \mathbb{F}_n^+} \|\theta_\alpha h\|^2 = \|\Theta h\|^2 \leq \|h\|^2 \end{aligned}$$

for any $h \in \mathcal{E}_u$. Therefore, we have equality and

$$\theta_\alpha | \mathcal{E}_u = 0 \quad \text{for any } \alpha \in \mathbb{F}_n^+, \ |\alpha| \geq 1.$$

Hence,

$$\Lambda := \theta | F^2(H_n) \otimes \mathcal{E}_u = I \otimes Z_u : F^2(H_n) \otimes \mathcal{E}_u \to F^2(H_n) \otimes \mathcal{E}_u'$$

is a unitary operator. Since Θ is a multi-analytic operator from $F^2(H_n) \otimes \mathcal{E}$ to $F^2(H_n) \otimes \mathcal{E}'$, we infer that

$$\Theta(F^2(H_n) \otimes \mathcal{E}_0) \subset F^2(H_n) \otimes \mathcal{E}_0'.$$

Hence, $\Psi := \Theta | F^2(H_n) \otimes \mathcal{E}_0$ is purely contractive and $\Theta = \Phi \oplus \Lambda$. The uniqueness part is straightforward, so we omit it. □

THEOREM 3.16. *Let $z_1, \ldots, z_k$ be distinct points in the unit ball $\mathbb{B}_n$, and let $C_j \in B(\mathcal{K}, \mathcal{K}')$, $j = 1, \ldots, k$. Let*

$$A := P_{\mathcal{N}'} R \in B(F^2(H_n) \otimes \mathcal{K}, \mathcal{N}'),$$

where where $R \in R_n^\infty \bar{\otimes} B(\mathcal{K}, \mathcal{K}')$ is an arbitrary operator such that $R(z_j) = C_j$, $j = 1, \ldots, k$ and the subspace $\mathcal{N}'$ is defined by relation (3.44). If $\dim \mathcal{K} < \infty$, then the central intertwining lifting $\Psi_{\max} \in R_n^\infty \bar{\otimes} B(\mathcal{K}, \mathcal{K}')$ of A is the maximal entropy solution for the Nevanlinna-Pick interpolation problem.

Let $m > k$ and $z_{k+1}, \ldots, z_m \in \mathbb{B}_n$ be such that $z_1, \ldots, z_k, z_{k+1}, \ldots, z_m$ are distinct points, and let

$$C_j := \Psi_{\max}(z_j), \quad j = k+1, \ldots, m.$$

Then $\Psi_{\max}$ is also the maximal entropy solution for the Nevanlinna-Pick inerpolation problem with the new data $\{z_j\}_{j=1}^m$ and $\{C_j\}_{j=1}^m$.

PROOF. Since $\mathcal{N}'$ is invariant under each operator $S_i^* \otimes I_{\mathcal{K}'}$, $i = 1, \dots, n$, we have

$$[F^2(H_n) \otimes \mathcal{K}'] \ominus \mathcal{N}' = \Phi(F^2(H_n) \otimes \mathcal{K}_1), \tag{3.45}$$

where $\Phi \in R_n^\infty \bar{\otimes} B(\mathcal{K}_1, \mathcal{K}')$ is an inner operator. Moreover, let us prove that Φ is a pure inner operator. According to Lemma 3.15, we have the decomposition $\Phi = \chi_1 \oplus \chi_2$ with the following properties:

(i) $\chi_1 \in R_n^\infty \bar{\otimes} B(\mathcal{E}_1, \mathcal{E}_1')$ is purely contractive;
(ii) $\chi_2 = I \otimes U$, where $U \in B(\mathcal{E}_2, \mathcal{E}_2')$ is a unitary operator;
(iii) $\mathcal{K}_1 = \mathcal{E}_1 \oplus \mathcal{E}_2$ and $\mathcal{K}' = \mathcal{E}_1' \oplus \mathcal{E}_2'$.

This implies that $F^2(H_n) \otimes \mathcal{E}_2'$ is in the range of Φ. According to relation (3.45), we have

$$F^2(H_n) \otimes \mathcal{E}_2' \subset F^2(H_n) \otimes \mathcal{K}'] \ominus \mathcal{N}'$$

and therefore $F^2(H_n) \otimes \mathcal{E}_2' \perp \mathcal{N}'$. Hence, for any $\alpha \in \mathbb{F}_n^+$, $k' \in \mathcal{K}'$, and $h \in \mathcal{E}_2'$, we have $e_\alpha \otimes h \perp f_{z_j} \otimes k'$. Hence, and taking into account the definition of f_{z_j}, we get

$$z_{i\alpha} \langle h, k' \rangle = 0 \tag{3.46}$$

for any $k' \in \mathcal{K}'$. Since $z_1, \dots, z_k$ are distinct points in the unit ball $\mathbb{B}_n$, we can find $z_{i\alpha} \neq 0$. Therefore, relation (3.46) implies $h = 0$. This proves that $\mathcal{E}_2' = \{0\}$, which shows that Φ is a pure inner operator.

Note that if $\Lambda \in R_n^\infty \bar{\otimes} B(\mathcal{K}, \mathcal{K}')$, then $P_{\mathcal{N}'}\Lambda = 0$ if and only if $\Lambda(z_j) = 0$ for any $j = 1, \dots, m$. Now, it is clear that $\Psi \in R_n^\infty \bar{\otimes} B(\mathcal{K}, \mathcal{K}')$ is a solution for the Nevanlinna-Pick interpolation problem with tolerance t if and only if Ψ is a solution for the problem (3.1), i.e.,

$$\Psi = R + \Theta G \quad \text{and} \quad \|\Psi\| \leq t,$$

where Φ is given by (3.45), and $R \in R_n^\infty \bar{\otimes} B(\mathcal{K}, \mathcal{K}')$ is such that $R(z_j) = C_j$, $j = 1, \dots, m$ (for the existence of such operator see [**4**]).

Using Theorem 3.1, we see that there is a solution for the Nevanlinna-Pick interpolation problem if and only if the operator $A := P_{\mathcal{N}'} R \in B(F^2(H_n) \otimes \mathcal{K}, \mathcal{N}')$ satisfies $\|A\| \leq t$. Moreover, the central intertwining lifting of A is the maximal entropy solution $\Psi_{\max}$ for the Nevanlinna-Pick interpolation problem.

Since the subspace

$$\mathcal{N}'' := \text{span}\,\{f_{z_j} : j = 1, \dots, m\} \otimes \mathcal{K}' \tag{3.47}$$

is invariant under each operator $S_i^* \otimes I_{\mathcal{K}'}$, $i = 1, \dots, n$, we have

$$[F^2(H_n) \otimes \mathcal{K}'] \ominus \mathcal{N}'' = \Psi(F^2(H_n) \otimes \mathcal{K}_2),$$

where $\Psi \in R_n^\infty \bar{\otimes} B(\mathcal{K}_2, \mathcal{K}')$ is an inner operator. Since $\mathcal{N}' \subset \mathcal{N}''$, we have

$$\Psi(F^2(H_n) \otimes \mathcal{K}_2) \subset \Phi(F^2(H_n) \otimes \mathcal{K}_1).$$

Hence, $\Psi = \Phi\Phi_1$ for some inner operator $\Psi_1 \in R_n^\infty \bar{\otimes} B(\mathcal{K}_2, \mathcal{K}_1)$ The Nevanlinna-Pick interpolation with data $\{z_j\}_{j=1}^m$ and $\{C_j\}_{j=1}^m$ is equivalent to the interpolation problem (3.4), i.e.,

$$\|\Gamma\| \leq t \quad \text{and} \quad \Gamma = \Psi_{\max} + \Phi\Phi_1 G.$$

Now, we can use the permanence principle of Theorem 3.2 to prove the last part of the theorem. □

Bibliography

[1] J. Agler and J. E. McCarthy, Complete Nevanlina-Pick kernels, *J. Funct. Anal.* **175** (2000), 111–124.

[2] A. Arias and G. Popescu, Factorization and reflexivity on Fock spaces, *Integr. Equat. Oper. Th.* **23** (1995), 268–286.

[3] A. Arias and G. Popescu, Noncommutative interpolation and Poisson transforms II, *Houston J. Math.* **25** (1999), 79–98.

[4] A. Arias and G. Popescu, Noncommutative interpolation and Poisson transforms, *Israel J. Math.* **115** (2000), 205–234.

[5] A. Z. Arov and M. G. Krein, On computations of entropy functionals and their minima, *Acta. Sci. Math. (Szeged)* **45** (1983), 51–66.

[6] W.B. Arveson, Subalgebras of C^*-algebras III: Multivariable operator theory, *Acta Math.* **181** (1998), 159–228.

[7] W.B. Arveson, The curvature invariant of a Hilbert module over $\mathbb{C}[z_1, \dots, z_n]$, *J. Reine Angew. Math.* **522** (2000), 173–236.

[8] J. A. Ball, T. T. Trent, and V. Vinikov, Interpolation and commutant lifting for multipliers on reproducing kernels Hilbert spaces, *Operator Theory and Analysis: The M.A. Kaashoek Anniversary Volume*, pages 89–138, **OT 122**, Birkhauser-Verlag, Basel-Boston-Berlin, 2001.

[9] J. A. Ball and V. Bolotnikov, On bitangential interpolation problem for contractive-valued functions on the unit ball, *Linear Algebra Appl.* **353** (2002), 107–147.

[10] J. W. Bunce, Models for n-tuples of noncommuting operators, *J. Funct. Anal.* **57**(1984), 21–30.

[11] J. Burg, *Maximum entropy spectral analysis*, Ph.D. dissertation, Stanford University, Stanford, 1975.

[12] C. Carathéodory, Über den Variabilitätsbereich der Koeffzienten von Potenzreinen die gegebene Werte nicht annehmen, *Math. Ann.* **64** (1907), 95–115.

[13] K. R. Davidson, E. Katsoulis, and D. Pitts, The structure of free semigroup algebras, *J. Reine Angew. Math.* **533** (2001), 99–125.

[14] K. R. Davidson and D. Pitts, Invariant subspaces and hyper-reflexivity for free semigroup algebras, *Proc. London Math. Soc.* **78** (1999), 401–430.

[15] K. R. Davidson and D. Pitts, Automorphisms and representations of the noncommutative analytic Toeplitz algebras, *Math. Ann.* **311** (1998), 275–303.

[16] K. R. Davidson and D. Pitts, Nevanlinna-Pick interpolation for noncommutative analytic Toeplitz algebras, *Integr. Equat. Oper. Th.* **31** (1998), 321–337.

[17] H. Dym and I. Gohberg, A maximum entropy principle for contractive interpolants, *J. Funct. Anal.* **65** (1986), 83–125.

[18] H. Dym and I. Gohberg, A new class of contractive interpolants and maximum entropy principle, *Topics in operator theory and interpolation*, 117–150, Oper. Theory Adv. Appl., 29, Birkhuser, Basel, 1988.

[19] R. L. Elis, I. Gohberg, and D. Lay, Band extensions maximum entropy and the permanence principle, *Maximum Entropy and Bayesian Methods in Applied Statistics*, Cambridge University Press, Cambridge, 1986.

[20] J. Eschmeier and M. Putinar, Spherical contractions and interpolation problems on the unit ball, *J. Reine Angew. Math.* **542** (2002), 219–236.

[21] C. Foias, A. E. Frazho, *The commutant lifting approach to interpolation problems*, Operator Theory: Advances and Applications, Birhäuser Verlag, Bassel, 1990.

[22] C. Foias, A. E. Frazho, and I. Gohberg, Central intertwining lifting, maximum entropy and their permanence, *Integr. Equat. Oper. Th.* **18** (1994), 166–201.

[23] C. Foias, A. E. Frazho, I. Gohberg, and M. Kaashoek, *Metric constrained interpolation, commutant lifting and systems*, Operator Theory: Advances and Applications vol. 100, Birhäuser Verlag, Bassel, 1998.

[24] A. E. Frazho, Models for noncommuting operators, *J. Funct. Anal.* **48**(1982), 1–11.

[25] B. A. Francis, *A Course in H^∞ Control Theory*, Lecture Notes in Control and Information Sciences, Springer-Verlag, New York, 1987.

[26] K. Glover and D. Mustafa, Derivation of the maximum entropy H_∞-controller and a state-space formula for its entropy, *Int. J. Cont.* **50** (1989), 899–916.

[27] K. Glover and D. Mustafa, *Minimum Entropy H_∞-control*, Lecture Notes in Control and Information Sciences, Springer-Verlag, New York, 1990.

[28] U. Grenander and G. Szegö, *Toeplitz forms and their applications*, Univ. of California Press, Berkeley, California, 1958.

[29] H. Helson and D. Lowdenslager, Prediction theory and Fourier series in several variables, *Acta Math.* **99** (1958), 165–202.

[30] H. Helson and D. Lowdenslager, Prediction theory and Fourier series in several variables II, *Acta Math.* **106** (1961), 175–213.

[31] K. Hoffman, *Banach Spaces of Analytic Functions*, Englewood Cliffs: Prentice-Hall, 1962.

[32] A. N. Kolmogorov, Sur l'interpolation et l'extrapolation des suites stationaire, *C. R. Acad. Sci. (Paris)* **208** (1939), 2043–2045.

[33] A. N. Kolmogorov , Stationary sequences in Hilbert spaces, *Bull. Math. Univ. Moscow* **2** (1941).

[34] Z. Nehari, On bounded bilinear forms, *Ann. of Math.* **65** (1957), 153–162.

[35] R. Nevanlinna, Über beschränkte Functionen, die in gegebenen Punkten vorgeschribene Werte annehmen, *Ann. Acad. Sci. Fenn. Ser A* **13** (1919), 7–23.

[36] V. I. Paulsen, G. Popescu, and D. Singh, On Bohr's inequality, *Proc. London Math. Soc.* **85** (2002), 493–512.

[37] G. Popescu, Models for infinite sequences of noncommuting operators, *Acta. Sci. Math. (Szeged)* **53**(1989), 355–368.

[38] G. Popescu, Isometric dilations for infinite sequences of noncommuting operators, *Trans. Amer. Math. Soc.* **316**(1989), 523–536.

[39] G. Popescu, Characteristic functions for infinite sequences of noncommuting operators, *J. Operator Theory* **22**(1989), 51–71.

[40] G. Popescu, Multi-analytic operators and some factorization theorems, *Indiana Univ. Math. J.* **38** (1989), 693–710.

[41] G. Popescu, On intertwining dilations for sequences of noncommuting operators, *J. Math. Anal. Appl.* **167** (1992), 382–402.

[42] G. Popescu, Von Neumann inequality for $(B(H)^n)_1$, *Math. Scand.* **68** (1991), 292–304.

[43] G. Popescu, Functional calculus for noncommuting operators, *Michigan Math. J.* **42** (1995), 345–356.

[44] G. Popescu, Multi-analytic operators on Fock spaces, *Math. Ann.* **303** (1995), 31–46.

[45] G. Popescu, Noncommutative disc algebras and their representations, *Proc. Amer. Math.* **124** (1996), 2137–2148.

[46] G. Popescu, Interpolation problems in several variables, *J. Math Anal. Appl.* **227** (1998), 227–250.

[47] G. Popescu, Poisson transforms on some C^*-algebras generated by isometries, *J. Funct. Anal.* **161** (1999), 27–61.

[48] G. Popescu, Structure and entropy for positive definite Toeplitz kernels on free semigroups, *J. Math. Anal. Appl.* **254** (2001), 191-218.

[49] G. Popescu, Spectral liftings in Banach algebras and interpolation in several variables, *Trans. Amer. Math. Soc.* **353** (2001), 2843–2857.

[50] G. Popescu, Commutant lifting, tensor algebras, and functional calculus, *Proc. Edinb. Math. Soc.* **44** (2001), 389-406.

[51] G. Popescu, Curvature invariant for Hilbert modules over free semigroup algebras, *Adv. Math.* **158** (2001), 264–309.

[52] G. Popescu, A lifting theorem for symmetric commutants, *Proc. Amer. Math.* **129** (2001), 1705–1711.

[53] G. POPESCU, Central intertwining lifting, suboptimization, and interpolation in several variables, *J. Funct. Anal.* **189** (2002), 132–154.

[54] G. POPESCU, Meromorphic interpolation in several variables, *Linear Algebra Appl.* **357** (2002), 173–196.

[55] G. POPESCU, Multivariable Nehari problem and interpolation, *J. Funct. Anal.* **200** (2003), 536–581.

[56] G. POPESCU, Similarity and ergodic theory of positive linear maps, *J. Reine Angew. Math.*, **561** (2003), 87–129.

[57] E. A. ROBINSON, *Random Wavelets and Cybernetic Systems*, Griffin, London, 1962.

[58] L. RODMAN , I. M. SPITKOVSKY, AND H. J. WOERDEMAN, Abstract band method via factorization, positive and band extensions of multivariable almost periodic matrix functions, and spectral estimation, *Mem. Amer. Math. Soc* **160** (2002), no. 762.

[59] M. ROSENBLUM AND J. ROVNYAK, *Hardy classes and operator theory*, Oxford University Press-New York, 1985.

[60] D. SARASON, Generalized interpolation in H^∞, *Trans. Amer. Math. Soc.* **127** (1967), 179–203.

[61] I. SCHUR, Über Potenzreihen die im innern des Einheitshreises beschränkt sind, *J. Reine Angew. Math.* **148** (1918), 122–145.

[62] G. SZEGÖ, Ein Grenzwertsvatz uber die Toeplitzschen Determinanten, einer reellen positiven Funktion, *Math. Ann.* **76** (1915), 490–503.

[63] G. SZEGÖ, Orthogonal polynomials, *Colloquium Publications* **23** (1939), Amer. Math. Soc. Providence, Rhode Island.

[64] G. SZEGÖ, *On certain Hermitian forms associated with the Fourier series of a positive function*, pp. 222–238, Festskrift Marcel Riesz, Lund, 1952.

[65] B. SZ.-NAGY AND C. FOIAŞ, Dilation des commutants d'opérateurs, *C. R. Acad. Sci. Paris, série A* **266** (1968), 493–495.

[66] B. SZ.-NAGY AND C. FOIAŞ, *Harmonic Analysis of Operators on Hilbert Space*, North Holland, New York 1970.

[67] N. WIENER, *Extrapolation, Interpolation, and Smoothing of stationary time series*, Wiley, New York, 1949.

[68] N. WIENER AND P. R. MASANI, 111–150 The prediction theory of multivariate stochastic processes I, *Acta Math.* **98** (1957), 111–150.

[69] N. WIENER AND P. R. MASANI, The prediction theory of multivariate stochastic processes II, *Acta Math.* **99** (1958), 93–139.

Editorial Information

To be published in the *Memoirs*, a paper must be correct, new, nontrivial, and significant. Further, it must be well written and of interest to a substantial number of mathematicians. Piecemeal results, such as an inconclusive step toward an unproved major theorem or a minor variation on a known result, are in general not acceptable for publication. Papers appearing in *Memoirs* are generally at least 80 and not more than 200 published pages in length. Papers less than 80 or more than 200 published pages require the approval of the Managing Editor of the Transactions/Memoirs Editorial Board.

As of July 31, 2006, the backlog for this journal was approximately 11 volumes. This estimate is the result of dividing the number of manuscripts for this journal in the Providence office that have not yet gone to the printer on the above date by the average number of monographs per volume over the previous twelve months, reduced by the number of volumes published in four months (the time necessary for preparing a volume for the printer). (There are 6 volumes per year, each containing at least 4 numbers.)

A Consent to Publish and Copyright Agreement is required before a paper will be published in the *Memoirs*. After a paper is accepted for publication, the Providence office will send a Consent to Publish and Copyright Agreement to all authors of the paper. By submitting a paper to the *Memoirs*, authors certify that the results have not been submitted to nor are they under consideration for publication by another journal, conference proceedings, or similar publication.

Information for Authors

Memoirs are printed from camera copy fully prepared by the author. This means that the finished book will look exactly like the copy submitted.

The paper must contain a *descriptive title* and an *abstract* that summarizes the article in language suitable for workers in the general field (algebra, analysis, etc.). The *descriptive title* should be short, but informative; useless or vague phrases such as "some remarks about" or "concerning" should be avoided. The *abstract* should be at least one complete sentence, and at most 300 words. Included with the footnotes to the paper should be the 2000 *Mathematics Subject Classification* representing the primary and secondary subjects of the article. The classifications are accessible from `www.ams.org/msc/`. The list of classifications is also available in print starting with the 1999 annual index of *Mathematical Reviews*. The Mathematics Subject Classification footnote may be followed by a list of *key words and phrases* describing the subject matter of the article and taken from it. Journal abbreviations used in bibliographies are listed in the latest *Mathematical Reviews* annual index. The series abbreviations are also accessible from `www.ams.org/publications/`. To help in preparing and verifying references, the AMS offers MR Lookup, a Reference Tool for Linking, at `www.ams.org/mrlookup/`. When the manuscript is submitted, authors should supply the editor with electronic addresses if available. These will be printed after the postal address at the end of the article.

Electronically prepared manuscripts. The AMS encourages electronically prepared manuscripts, with a strong preference for $\mathcal{A}_{\mathcal{M}}\mathcal{S}$-LaTeX. To this end, the Society has prepared $\mathcal{A}_{\mathcal{M}}\mathcal{S}$-LaTeX author packages for each AMS publication. Author packages include instructions for preparing electronic manuscripts, the *AMS Author Handbook*, samples, and a style file that generates the particular design specifications of that publication series. Though $\mathcal{A}_{\mathcal{M}}\mathcal{S}$-LaTeX is the highly preferred format of TeX, author packages are also available in $\mathcal{A}_{\mathcal{M}}\mathcal{S}$-TeX.

Authors may retrieve an author package from e-MATH starting from `www.ams.org/tex/` or via FTP to `ftp.ams.org` (login as `anonymous`, enter username as password, and type `cd pub/author-info`). The *AMS Author Handbook* and the *Instruction Manual* are available in PDF format following the author packages link from `www.ams.org/tex/`. The author package can also be obtained free of charge by sending email to tech-support@ams.org (Internet) or from the Publication Division, American Mathematical Society, 201 Charles St., Providence, RI 02904-2294, USA. When requesting an author package, please specify $\mathcal{A}\mathcal{M}\mathcal{S}$-LaTeX or $\mathcal{A}\mathcal{M}\mathcal{S}$-TeX and the publication in which your paper will appear. Please be sure to include your complete mailing address.

Sending electronic files. After acceptance, the source file(s) should be sent to the Providence office (this includes any TeX source file, any graphics files, and the DVI or PostScript file).

Before sending the source file, be sure you have proofread your paper carefully. The files you send must be the EXACT files used to generate the proof copy that was accepted for publication. For all publications, authors are required to send a printed copy of their paper, which exactly matches the copy approved for publication, along with any graphics that will appear in the paper.

TeX files may be submitted by email, FTP, or on diskette. The DVI file(s) and PostScript files should be submitted only by FTP or on diskette unless they are encoded properly to submit through email. (DVI files are binary and PostScript files tend to be very large.)

Electronically prepared manuscripts can be sent via email to pub-submit@ams.org (Internet). The subject line of the message should include the publication code to identify it as a Memoir. TeX source files, DVI files, and PostScript files can be transferred over the Internet by FTP to the Internet node e-math.ams.org (130.44.1.100).

Electronic graphics. Comprehensive instructions on preparing graphics are available at www.ams.org/jourhtml/graphics.html. A few of the major requirements are given here.

Submit files for graphics as EPS (Encapsulated PostScript) files. This includes graphics originated via a graphics application as well as scanned photographs or other computer-generated images. If this is not possible, TIFF files are acceptable as long as they can be opened in Adobe Photoshop or Illustrator. No matter what method was used to produce the graphic, it is necessary to provide a paper copy to the AMS.

Authors using graphics packages for the creation of electronic art should also avoid the use of any lines thinner than 0.5 points in width. Many graphics packages allow the user to specify a "hairline" for a very thin line. Hairlines often look acceptable when proofed on a typical laser printer. However, when produced on a high-resolution laser imagesetter, hairlines become nearly invisible and will be lost entirely in the final printing process.

Screens should be set to values between 15% and 85%. Screens which fall outside of this range are too light or too dark to print correctly. Variations of screens within a graphic should be no less than 10%.

Inquiries. Any inquiries concerning a paper that has been accepted for publication should be sent directly to the Electronic Prepress Department, American Mathematical Society, 201 Charles St., Providence, RI 02904, USA.

Editors

This journal is designed particularly for long research papers, normally at least 80 pages in length, and groups of cognate papers in pure and applied mathematics. Papers intended for publication in the *Memoirs* should be addressed to one of the following editors. In principle the Memoirs welcomes electronic submissions, and some of the editors, those whose names appear below with an asterisk (*), have indicated that they prefer them. However, editors reserve the right to request hard copies after papers have been submitted electronically. Authors are advised to make preliminary email inquiries to editors about whether they are likely to be able to handle submissions in a particular electronic form.

Titles in This Series

868 **Gelu Popescu,** Entropy and multivariable interpolation, 2006

867 **Vilmos Totik,** Metric properties of harmonic measures, 2006

866 **William Craig,** Semigroups underlying first-order logic, 2006

865 **Nathanial P. Brown,** Invariant means and finite representation theory of C^*-algebras, 2006

864 **John M. Lee,** Fredholm operators and Einstein metrics on conformally compact manifolds, 2006

863 **M. Lübke and A. Teleman,** The Universal Kobayashi-Hitchin correspondence on Hermitian manifolds, 2006

862 **Alberto Canonaco,** The Beilinson complex and canonical rings of irregular surfaces, 2006

861 **Leon A. Takhtajan and Lee-Peng Teo,** Weil-Petersson metric on the universal Teichmüller space, 2006

860 **Thomas M. Fiore,** Pseudo limits, biadjoints and pseudo algebras: Categorical foundations of conformal field theory, 2006

859 **N. Arcozzi, R. Rochberg, and E. Sawyer,** Carleson measures and interpolating sequences for Besov spaces on complex balls, 2006

858 **Enrico Valdinoci, Berardino Sciunzi, and Vasile Ovidiu Savin,** Flat level set regularity of p-Laplace phase transitions, 2006

857 **Donatella Danielli, Nocola Garofalo, and Duy-Minh Nhieu,** Non-doubling Ahlfors measures, perimeter measures, and the characterization of the trace spaces of Sobolev functions in Carnot-Carathéodory spaces, 2006

856 **Vladimir Bolotnikov and Harry Dym,** On boundary interpolation for matrix valued Schur functions, 2006

855 **Yevgenia Kashina, Yorck Sommerhäuser, and Yongchang Zhu,** On higher Frobenius-Schur indicators, 2006

854 **Noam Greenberg,** The role of true finiteness in the admissible recursively enumerable degrees, 2006

853 **Joachim Krieger,** Stability of spherically symmetric wave maps, 2006

852 **Viorel Barbu, Irena Lasiecka, and Roberto Triggiani,** Tangential boundary stabilization of Navier-Stokes equations, 2006

851 **Jie Wu,** On maps from loop suspensions to loop spaces and the shuffle relations on the Cohen groups, 2006

850 **Siegfried Echterhoff, S. Kaliszewski, John Quigg, and Iain Raeburn,** A categorical approach to imprimitivity theorems for C^*-dynamical systems, 2006

849 **Katsuhiko Kuribayashi, Mamoru Mimura, and Tetsu Nishimoto,** Twisted tensor products related to the cohomology of the classifying spaces of loop groups, 2006

848 **Bob Oliver,** Equivalences of classifying spaces completed at the prime two, 2006

847 **Eric T. Sawyer and Richard L. Wheeden,** Hölder continuity of weak solutions to subelliptic equations with rough coefficients, 2006

846 **Victor Beresnevich, Detta Dickinson, and Sanju Velani,** Measure theoretic laws for lim–sup sets, 2006

845 **Ehud Friedgut, Vojtech Rödl, Andrzej Ruciński, and Prasad V. Tetali,** A Sharp threshold for random graphs with a monochromatic triangle in every edge coloring, 2006

844 **Amadeu Delshams, Rafael de la Llave, and Tere M. Seara,** A geometric mechanism for diffusion in Hamiltonian systems overcoming the large gap problem: Heuristics and rigorous verification on a model, 2006

843 **Denis V. Osin,** Relatively hyperbolic groups: Intrinsic geometry, algebraic properties, and algorithmic problems, 2006

842 **David P. Blecher and Vrej Zarikian,** The calculus of one-sided M-ideals and multipliers in operator spaces, 2006

For a complete list of titles in this series, visit the AMS Bookstore at **www.ams.org/bookstore/**.